CAMBRIDGE LIBRARY COLLECTION

Books of enduring scholarly value

Life Sciences

Until the nineteenth century, the various subjects now known as the life sciences were regarded either as arcane studies which had little impact on ordinary daily life, or as a genteel hobby for the leisured classes. The increasing academic rigour and systematisation brought to the study of botany, zoology and other disciplines, and their adoption in university curricula, are reflected in the books reissued in this series.

The Life of Sir Joseph Banks

Sir Joseph Banks (1743–1820), botanist and patron of science, was a pivotal figure in eighteenth-century intellectual circles. He travelled around the world with Captain Cook as naturalist on the *Endeavour* (1768–1771), exploring first Tahiti, then Australia, New Zealand and Indonesia, and contributed £10,000 of his personal wealth to help finance the expedition. He became President of the Royal Society and scientific adviser to the Royal Gardens at Kew, counting George III as a personal friend. He both helped plan the first penal colony in New South Wales, and bred Merino sheep to be farmed there. He promoted the geological mapping of England, Flinders' circumnavigation of Australia, and the transfer of breadfruit from the Pacific to the West Indies (the objective of the *Bounty* voyage that ended in mutiny). This 1911 study, based on extensive archival research, was the first detailed biography of this remarkable and influential man.

Cambridge University Press has long been a pioneer in the reissuing of out-of-print titles from its own backlist, producing digital reprints of books that are still sought after by scholars and students but could not be reprinted economically using traditional technology. The Cambridge Library Collection extends this activity to a wider range of books which are still of importance to researchers and professionals, either for the source material they contain, or as landmarks in the history of their academic discipline.

Drawing from the world-renowned collections in the Cambridge University Library, and guided by the advice of experts in each subject area, Cambridge University Press is using state-of-the-art scanning machines in its own Printing House to capture the content of each book selected for inclusion. The files are processed to give a consistently clear, crisp image, and the books finished to the high quality standard for which the Press is recognised around the world. The latest print-on-demand technology ensures that the books will remain available indefinitely, and that orders for single or multiple copies can quickly be supplied.

The Cambridge Library Collection will bring back to life books of enduring scholarly value (including out-of-copyright works originally issued by other publishers) across a wide range of disciplines in the humanities and social sciences and in science and technology.

The Life of
Sir Joseph Banks

Edward Smith

CAMBRIDGE
UNIVERSITY PRESS

CAMBRIDGE UNIVERSITY PRESS

Cambridge, New York, Melbourne, Madrid, Cape Town,
Singapore, São Paolo, Delhi, Tokyo, Mexico City

Published in the United States of America by Cambridge University Press, New York

www.cambridge.org
Information on this title: www.cambridge.org/9781108031127

© in this compilation Cambridge University Press 2011

This edition first published 1911
This digitally printed version 2011

ISBN 978-1-108-03112-7 Paperback

: : THE LIFE OF : :
SIR JOSEPH BANKS

Sir Joseph Banks, K.B., P.R.S.

From a portrait by Sir Thomas Lawrence, P.R.A.

:: THE LIFE OF ::
SIR JOSEPH BANKS

PRESIDENT OF THE ROYAL SOCIETY
WITH SOME NOTICES OF HIS
FRIENDS AND CONTEMPORARIES
❦ BY EDWARD SMITH, F.R.H.S. ❦
WITH A PHOTOGRAVURE FRONTISPIECE
AND SIXTEEN OTHER ILLUSTRATIONS

LONDON : JOHN LANE, THE BODLEY HEAD
NEW YORK: JOHN LANE COMPANY. MCMXI

ERRATA.

Illustration facing page 314.   For George Cruickshank read John Gillray.

Page XVI, last line.   For G. Cruickshank read John Gillray.

DEDICATED
BY PERMISSION
TO
SIR JOSEPH DALTON HOOKER, O.M., G.C.S.I., M.D., F.R.S.

PREFACE

WITH the idea of presenting to modern readers an unfamiliar side of the eighteenth century, in which Science and Public Spirit would be represented, it appeared to me that a prominent figure could be chosen to serve as a common centre. The eye fell upon the name of Sir Joseph Banks : concerning whom many persons have vague notions of a distinguished character, not yet unveiled to the student of English social history.

Banks's record is that of a man of unbounded Public Spirit. Science was his passion, and the public service through the applications of science was his constant aim. Fortune favoured him in several ways, and he left an enduring mark upon the times in which he lived. The ensuing pages will be something of a revelation ; especially in the devotion and regard shown to Banks by so many varied characters among his cotemporaries. The man has practically vanished from our ken, as an individual ; but, as the tale develops, we shall discover that Banks was the inspiring agent of a number of useful works which have permanently benefited the world.

A detailed life of this worthy man has always been wanted. Several abortive attempts were made to collect his papers with the view to an authoritative biography. When another generation arose which knew little of him personally, the opportunity had passed for reproducing those intimacies which only surviving friends could deal with. Thus the personality of a man who had held a foremost place in scientific society and in the polite world for upwards of forty years was lost. The grateful recollection of friends and cotemporaries was soon only to be found in some stray memorial sketches.[1]

There were many noble examples of men, during the eighteenth century, who left an impress of goodness and greatness upon their time. Banks was a child of Fortune, favoured alike in his manly and generous person and in the possession of ample means wherewith to gratify his liberal tastes. Light has been thrown upon many fine characters, hitherto almost forgotten, through the *Dictionary of National Biography*. In that useful cyclopædia of the men who have made Britain famous, we are ever reminded of the truth that *Man is never so great as his work*. In the case before us, our own generation knows him not; but the fruit of his labours is ever with us.

Sir Joseph Banks gave enormous impulse to the study of Natural Science, and to the improvement of social conditions.

His conduct of the Royal Society for forty years is,

---

[1] Since this work was finished, a book has reached London from New South Wales : an excellent and deeply interesting record of one side of Banks's career, under the title of *Sir Joseph Banks, the Father of Australia*, by J. H. Maiden, F.L.S., Director of the Sydney Botanic Garden. Mr. Maiden's book has had great success in Australia ; which is not wonderful, seeing that the people of New South Wales have kept green the memory of Banks, and still hold his name in undying honour.

perhaps, the most notable feature of his career. To this are to be added so many public functions that, as will be seen, there was scarcely any important movement in which he had not an active share. The Society of Arts, the Society of Antiquaries, the Linnean Society, the Royal Horticultural Society, and a host of local institutions of similar character all over Europe, found in him an active friend. They still continue to hold his name in veneration. There were enthusiastic naturalists of his day in France and Germany, Sweden, Italy, and Russia, who held Banks to be the greatest living Englishman. And no wonder. He had shared the famous circumnavigation with Captain Cook; he subsidized botanists and explorers all over the world; his natural history collections were at the service of everybody; he made Kew the botanical Mecca. It was Banks who inspired the famous adventure of Captain Bligh for transplanting the breadfruit, and ultimately made it successful. The botanic gardens in India and our Colonies have traditions of Banks that they will never lose.

The papers and journals of Banks have been long since dispersed. But the British Museum has rescued from autograph-hunters and others a great number of letters addressed to him. At the Natural History Museum is a collection of some two thousand copies of letters and semi-official documents, which were written under the care of Mr. Dawson Turner. Beside these, there are in the custody of the Director at Kew several hundred letters, chiefly from Banks's botanical and scientific friends. Upon all these, I have been able to trace at least a fair outline of Sir Joseph's career. I have to thank most cordially the gentlemen officially connected with these collections, for their assistance, and for their enthusiastic countenance to the plan for raising a memorial in literary shape to the man they delight to honour. Banks's library still occupies a spacious room in Montague House. His botanical specimens are dispersed among the treasures at the Natural History Museum.

Sir Joseph Banks made no pretensions to authorship. Beyond a few items, mentioned in the text, he seldom appeared in print during his lifetime. Yet, it would be difficult to find many persons who wrote so much and so readily. It is clear that he was a copious diarist; for he has always his dates and his facts at hand when required. Some of his statements, or reports, or instructions, that happen to be extant are brimful of information; of a sort which could only be provided from his own personal records. He would always answer a letter, if possible, which had come from the most insignificant correspondent. There was generally an amanuensis at hand; and for the very good reason that his handwriting was "shocking." Those of his papers which the present writer has had an opportunity of perusing have been generally rough drafts, and they are not easy to read. The immense accumulation of copies—there are twenty volumes of them—now in the custody of the Natural History Museum, manifest in frequent hiatus the difficulties of the copyist. All this goes further to prove the extraordinary fullness of the man's life.

The diary habit of Banks is further shown in his fragments of tours and travel-notes. These remain mostly unpublished. Only one has been done justice to, at the hands of Sir J. D. Hooker, as mentioned in the course of our narrative.

But his name occurs in innumerable " Dedications." He assisted freely in the publications of other students of Botany. And four books, at least, were professedly edited by him, and issued at his own expense. These are :

(1) *Reliquiæ Houstoniæ* (London, 1781) ; consisting of notes on the herbarium of William Houston, a naval surgeon, who brought many rare plants from Mexico and the West Indies.

(2) *Icones Plantarum quæ in Japonia collegit et delineavit Ingelbert Kæmpfer.* (London, 1791.)

(3) *Plants of the Coast of Coromandel* . . . [by W. Roxburgh]. Published under the direction of Sir Joseph Banks. (London, 1795–1819. 3 vols. folio.)

(4) *Practical Observations on the British Grasses* . . . by William Curtis. Fourth edition, to which is now added a short account of the diseases in corn. By Sir Joseph Banks. (London, 1805, 8vo.)

To these should be added the marvellous catalogue of his Library, prepared by Jonas Dryander : *Catalogus bibliothecæ historico-naturalia* . . . *Josephi Banks Baroneti.* (5 vols. Lond. 1798–1800.)

.   .   .   .   .

The author is indebted to his friend Dr. B. Daydon Jackson, Secretary and Librarian of the Linnean Society, for kindly reading the proofs of this volume ; and for permitting the use of two important pictures belonging to the Society. His thanks are also due to the Honourable Mrs. Knatchbull-Hugessen, who has generously placed at the disposal of the publisher the four beautiful drawings by John Russell, R.A., which add such conspicuous value to the group of illustrations with which he has adorned the work. Another "good genius" (from an author's point of view) is Gery Milner-Gibson Cullum, Esq., of Hardwick Hall, who was so kind as to allow me to explore the unedited correspondence of Sir John Cullum and his friends.

There has not been space to do full justice, within the limits of this volume, to the mass of material which has passed under the author's eye. But he ventures to presume so far as to say that he has made out a case for fuller study of the under-currents of life during the despised eighteenth century. Beside and beyond the witches' cauldron of frivolity and dishonour which is so often presented to us, there is evidently an unknown background of sterling virtues, of greatness, virility, moral rectitude, and the love of wisdom for its own sake.

E. S.

CONTENTS

CHAPTER I

PAGE

EARLY YEARS 3

CHAPTER II

ROUND THE WORLD WITH CAPTAIN COOK . . . 14

CHAPTER III

VISIT TO ICELAND. COOK'S SECOND AND THIRD VOYAGES 32

CHAPTER IV

PRESIDENT OF THE ROYAL SOCIETY 55

CHAPTER V

THE ROYAL SOCIETY—continued 69

CHAPTER VI

KEW GARDENS—GEORGE III 93

CHAPTER VII

PLANT COLLECTORS 111

CHAPTER VIII

BLIGH'S VOYAGES IN THE "BOUNTY" AND "PROVIDENCE" 126

CHAPTER IX

VARIOUS ADVENTURERS 139

CHAPTER X

MÆCENAS AND HIS HAPPENINGS 156

CHAPTER XI

THE SCOFFER ABROAD   172

CHAPTER XII

EUROPEAN FAME   192

CHAPTER XIII

THE FOUNDING OF AUSTRALIA   .   .   . 213

CHAPTER XIV

CAPTAIN FLINDERS AND ROBERT BROWN   .   .   . 230

CHAPTER XV

ICELANDIC AFFAIRS 244

CHAPTER XVI

THE RISE OF NEW LEARNED SOCIETIES   .   .   . 255

CHAPTER XVII

REVIVAL OF BOTANICAL EXPLORATION   .   .   . 266

CHAPTER XVIII

FAILING HEALTH, BUT UNFLAGGING ZEAL   .   .   . 280

CHAPTER XIX

SOME FRIENDS OF LATER YEARS   290

CHAPTER XX

"A FINE OLD ENGLISH GENTLEMAN"   .   .   . 304

CHAPTER XXI

THE END 322

ILLUSTRATIONS

SIR JOSEPH BANKS . . . *Frontispiece*
From a painting by Sir Thomas Lawrence, in the British Museum.

PAGE

MRS. WILLIAM BANKS 6
From a drawing by John Russell, R.A. In Lady Brabourne's collection.
(Reproduced from a photograph in the possession of the Hon. Mrs.
Knatchbull-Hugessen.)

JOSEPH BANKS 30
From a mezzotint by Dickinson, after Sir J. Reynolds.

OMAI THE OTAHEITAN 42
From a mezzotint by Jacobi, after Sir J. Reynolds.

SIR JOSEPH AND LADY BANKS 62
From a Wedgwood cameo, attributed to Flaxman, at the Linnean Society.

SPRING GROVE, IN HESTON, MIDDLESEX, Residence of Sir
Joseph Banks 90
From "Beauties of England and Wales."

SIR JOSEPH BANKS 156
From a drawing by John Russell, R.A. In Lady Brabourne's collection.
(Reproduced from a photograph in the possession of the Hon. Mrs.
Knatchbull-Hugessen.)

LADY BANKS 158
From a drawing by John Russell, R.A. In Lady Brabourne's collection.
(Reproduced from a photograph in the possession of the Hon. Mrs.
Knatchbull-Hugessen.)

MACARONIS 176
From an old print.

THE FLY CLUB 178
From a caricature by T. Rowlandson.

THE BATH BUTTERFLY 180
From a caricature by J. Gillray.

xvi   ILLUSTRATIONS

                                                        PAGE
ROBERT BROWN, F.L.S. 242
    From a portrait by H. W. Pickersgill, R.A., belonging to the Linnean Society.

LANDING THE TREASURES 286
    From a caricature by G. Cruikshank, after a design of Capt. Marryat.

REVESBY ABBEY 304
    From Howlett's Views in the County of Lincoln.

THE ANTIQUARIAN SOCIETY 314
    From a caricature by G. Cruikshank.

MISTRESS SARAH SOPHIA BANKS 322
    From a drawing by John Russell, R.A. In Lady Brabourne's collection.
    (Reproduced from a photograph in the possession of the Hon. Mrs.
    Knatchbull-Hugessen.)

AN OLD MAID ON A JOURNEY 324
    From a caricature by G. Cruikshank.

: : THE LIFE OF : :
SIR JOSEPH BANKS
PRESIDENT OF THE ROYAL SOCIETY

: : THE LIFE OF : :
SIR JOSEPH BANKS

CHAPTER I

EARLY YEARS

JOSEPH BANKS was born in Argyle Street, London, February, 1743.

His great-grandfather, Joseph Banks, was representative of a family of some wealth and position in Lincolnshire ; member of Parliament for Grimsby, and later for Totnes. He was an enthusiastic antiquary ; and had for friends and correspondents such men as Browne Willis, Doctor William Stukeley, and others. Warmly attached to his house of Revesby Abbey, he made his home there ; and probably travelled little, except to London on his parliamentary duties. In a letter to Stukeley, he speaks gratefully of a friend who writes to tell him how the world goes on, news so welcome in a remote place like his. He wishes the Doctor was able to be with him: for plenty of unworldly news was within reach :

" I would show you a sight of eight Religious Houses, very great ones, in twelve miles riding, in the nearest road from my house to Lincoln, all within two hundred paces of the road. Revesby Abbey, Tattershall College, Kirkstead, Stickswould, Axholme, Bardney, Stanfield and Barlings abbeys ; which show you what fine folks we have been

formerly. . . ." Browne Willis found Banks thirsting for
information on the history of his county, and worked up
some records of Revesby for him. And he gave the profound advice for every gentleman's son—" every lad of a
mannour"—to collect in a book what he could of his parish,
urging his old friend to employ a transcriber from the
Records, when next time he is called up to London.

Mr. Banks died in 1727, aged sixty-two years. His
son Joseph was another honoured squire : Fellow of the
Society of Antiquaries, and a member of the Spalding
Gentlemen's Society. He was member of Parliament for
Peterborough, and sometime Sheriff of Lincolnshire. He
rebuilt Revesby Church.

The next generation is represented by William Banks,
who succeeded to the family estates on his father's death
in 1736. This was another public-spirited man, doubtless
a person of some consequence in the county. There was
published, in 1749, an " Open Letter " to this gentleman
on " Distemper among Horned Cattle " : a circumstance
which would imply something of leadership in agricultural
improvement. He represented Grampound in Parliament.

William Banks's wife was Marianne, daughter of William
Bate, Esq. One would like to know more of this lady,
the parent of such a masterful character as was that
of Sir Joseph throughout his long life. Nowadays, we
justly insist upon the share of mental and moral vigour in
a mother, as a factor in the mental and moral development of a great and good man. The little that is recorded
of Mrs. William Banks, however, goes far enough to show
that she encouraged instead of thwarted the instincts
which afterward led her son to deserved honour. And
she had the great blessing of witnessing the best part of
his career. She died at an advanced age in 1804, at Sir
Joseph's house in Soho Square.[1]

[1] " A lady remarkable for her charities and piety ; and devoted to
her religious duties in the Church of England."—G. Suttor.

After suitable preparation at home, young Banks was placed at Harrow School, where he spent about four years. At thirteen he was sent to Eton. The master soon recognized a cheerful and generous disposition, but it was disappointing that he did not promise to be a scholar. He was obviously an open-air boy ; fond of sport and play ; not incapable of attentive study, yet not disposed to it in a " bookish " sense. Activity and energy in out-of-door pursuits was his most characteristic feature. This, of course, made for honour with his school-fellows, with whom he speedily became very popular.

One of his chosen associates was Mr. Brougham, father of the future Chancellor. They were of the same age ; were both fond of long walks, and expert in swimming. " My father described him as a remarkably fine-looking, strong, and active boy, whom no fatigue could subdue, and no peril daunt ; and his whole time out of school was given up to hunting after plants and insects, making a *hortus siccus* of the one, and forming a cabinet of the other. As often as Banks could induce him to quit his task in reading or in verse-making, he would take him on his long rambles ; and I suppose it was from this early taste that we had at Brougham so many butterflies, beetles, and other insects, as well as a cabinet of shells and fossils." [1]

This taste for Natural Science was doubtless innate, only awakened by the accidents of his environment. The circumstance which somewhat altered his habits and made him a keen observer of nature was related by Banks in after years to his friend Sir Everard Home. [2] One day he had been bathing with his fellow Etonians ; and on coming out of the water to dress he found that all but himself had gone away. Having put on his clothes, he walked slowly along a green lane. It was a fine summer's

[1] Brougham : *Lives of Men of Letters and Science Who Flourished in the Time of George III.*
[2] *Hunterian oration.*

evening ; flowers covered the sides of the path. He felt
delighted with the natural beauties around him, and
exclaimed, " How beautiful ! would it not be far more
reasonable to make me learn the nature of these plants
than the Greek and Latin I am confined to ? " His next
reflection was that he must do his duty, obey his father's
commands, and reconcile himself to the learning of the
school. But this did not hinder him from the study of
Botany ; and having no better instructor he paid some
women who were employed in gathering plants for the
druggists, for such information as they could give him.
Returning home for the holidays, he was inexpressibly
delighted to find in his mother's dressing-room an old
copy of Gerarde's *Herbal*, having the names and figures
of those plants with which he had formed an imperfect
acquaintance, and he carried it with him back to school.
There he continued his collection of plants, and he also
made one of butterflies and other insects.

After leaving school, young Banks was entered at
Christ Church, Oxford. Meanwhile, early in the year 1761,
Mr. William Banks died, leaving Joseph and a sister to
the care of his widow. For the present Mrs. Banks left
Revesby for London, and took up her residence at Chelsea
with her children. She had learned that there were
peculiar advantages here. It was healthy, and it was
cheerful ; and they could live in the proximity of numerous gardens. Her house was a Queen Anne mansion,
situated in Paradise Walk, near the famous Apothecaries'
Garden. She was determined to gratify her boy's bent
for Botany and Natural History.

These were the palmy days of Chelsea. It had its own
society. The houses of the aristocracy, with fine gardens,
were numerous ; and they abounded with companies of
well-cultured people. The Thames is wider here than at
any other point above London Bridge, and in those days,
if we may trust all the pleasant memories that have been

MRS. WILLIAM BANKS
From a drawing by John Russell, R.A.

recorded of it, Chelsea must have been a very charming suburb. All these things suggest a happy combination of circumstances ; under which young Banks was able to indulge in his favourite pursuits without losing any social advantages.

A story is related of Banks which perhaps belongs to this period. He was out one day searching for plants, on the bank of a ditch near Hounslow. A gentleman in a post-chaise had just been robbed by a highwayman. After the latter had decamped, the traveller proposed to his driver that he should take the horses out of the carriage and go in pursuit of the villain. They had not gone far when they saw Mr. Banks under a hedge searching for a plant, with the bridle of his horse upon his arm. The post-boy called out, " Here he is ! " Banks was forthwith conveyed, in spite of protestations, to Bow Street, and charged before Sir John Fielding with the robbery. The business ended with apologies and regrets, as soon as the prisoner had satisfied the magistrate as to his identity.

Banks presently departed for Oxford. From the time he entered his college, it became clear that Fate had determined for him a career far removed from the traditional ambitions of the University. Scholarship did not attract him. Nor did he, although a social favourite, yield to the fascinations of any set of mere pleasure-loving undergraduates. He found a professorship of Botany in existence. All other considerations yielded to this ; and he at once made personal advances to the holder.

This was Dr. Humphrey Sibthorp, of whom it was said that he only gave one lecture in the course of thirty-five years. He was father of John Sibthorp, a man who came to higher distinction in botanical science, travelled widely in search of plants, and was one of the founders of the Linnean Society. Forthwith went our young enthusiast to Professor Sibthorp, and was happy in arousing his interest. Banks proposed to secure the services of a reader or lecturer in Botany, whose remuneration should be met by subscription from his pupils. With very creditable kindness, Dr. Sibthorp at once acceded. There was no one at Oxford willing or able to undertake the thing. Banks, therefore, rode over to Cambridge in search of a candidate ; made acquaintance with Dr. Martyn, the Botanical Professor, and speedily found the very man that was wanted. This was Israel Lyons, a young man about four years the senior of Banks. He was clever both in botany and mathematics. His father was a noted character in the town of Cambridge, where he kept a silversmith's shop, and taught Hebrew to some of the University students. The younger Israel Lyons was a remarkable specimen of the numerous class of self-educated persons who helped to make the eighteenth century. Like Banks, he had acquired proficiency in Botany during his boyhood. He found a patron in Dr. Robert Smith, master of Trinity, who was greatly impressed by his mathematical knowledge. Lyons had much promise of distinction.[1]

The experiment at Oxford was justified by events. A few students gathered round, and Banks was rewarded by seeing a new taste for Natural History come into favour. As long as he remained at Oxford, Lyons was encouraged by his pupil's progress in botanical knowledge, and by the distinction which his example gave to the study of the science.

Banks left Oxford in December, 1763. In the following February he was of age, and then entered into the possession of his ancestral property. He lived much at Revesby, with his mother and sister. He had no disposition to

---

[1] He went as astronomer with Captain Phipps's Expedition toward the North Pole (1773). He presently married well, and settled in London. But he ended badly, and did not live to reach middle age.

idleness. The pursuit of Natural History was followed with the greatest ardour ; " his relaxation was confined to exercise, and to angling, of which he was so fond that he would devote days and nights to it ; and as it happened that Lord Sandwich had the same taste, and both possessed estates in Lincolnshire, they became intimately acquainted, and saw much of each other. So zealous were these friends in the prosecution of this sport, that they formed a project for suddenly draining the Serpentine by letting off the water. Their hope was to have thrown much light on the state and habits of the fish. Banks was wont to lament their scheme being discovered the night before it was to have been executed " (Brougham).

A notable early acquaintance of Banks was Thomas Pennant. He was a good zoologist, and produced several popular books on birds and other fauna ; although there was difference of opinion among his friends as to the value of his attainments in natural history. He had a better reputation for an observing traveller in his own country : one that is still deserving of record for those who would get a glimpse of Britain in Pennant's days. His seat at Downing, Flintshire, was the centre of an intelligent and very wide circle of friends. Banks and Pennant had more than one botanizing tour together.

Another botanical friend was the Rev. John Lightfoot, who later acquired some distinction in the science. Such men as these entered into Banks's life in these earlier days. Yet he did not keep outside the great world of London. He had a town house in New Burlington Street. Here he attracted the society of intelligent persons, especially those who were able to collaborate in the study of Natural Science. There must have been something attractive about his personality as a young man. His circle of congenial friends was ever widening. Nothing could be more suggestive of his social and in-

tellectual status than his election to the Royal Society
at the early age of twenty-three. Banks was nominated
(April, 1766) by Dr. Charles Lyttleton, Bishop of Carlisle,
an antiquary of the period ; Dr. Morton, librarian at the
British Museum ; William Watson, M.D., who always remained a friend and associate of Banks, and was himself
highly distinguished in physics and astronomy ; Richard
Kaye, afterwards Dean of Lincoln ; and Mr. James West,
a former President of the Society. Banks was already
a member of the Society of Arts, and in time came to be
one of its most assiduous supporters.

Banks had what were then unusually wide views as to
the possibility of extending his knowledge of botany and
natural history. It was nothing new to send out explorers and collectors of exotic plants. But he was,
perhaps, the first young man of fortune who was induced
personally to take the hazards of such a life. In this very
year, 1766, an opportunity occurred for a first adventure,
or what his friend Lightfoot would call a freak, of this
sort.

The occasion was this. The *Niger*, Captain Thomas
Adams, was ordered to Newfoundland and Labrador on
business concerning the fisheries. An Oxford friend of
Banks, Lieutenant Constantine Phipps, was on board.
Phipps was at the beginning of a career of some little
note. After several years of naval life, he succeeded his
father as Lord Mulgrave, and took an active part in
public affairs. Now, at twenty-two years of age, he was
prepared for any step which would lead to distinction.
The spirit of adventure was in the air, and specially
attractive was any project for exploring distant lands.
The idea of Banks accompanying the ship with the view
of studying Natural History was made only to be warmly
adopted. Preparations for joining the *Niger* were made
with that profusion, and disregard of expense, which
characterized Banks during all his lifetime when there

was a worthy object in view, and when there was a
scientific trip to the fore.

He left London on April 7, 1766, and joined his
friend Phipps at the latter's house in Hampshire. Before
leaving Plymouth they visited Mount Edgcumbe, Banks
botanizing there as, indeed, he had done all along the
route, beside taking intelligent notice of other matters.
The *Niger* sailed April 22, and reached Newfoundland
May 11. On landing, an active life began at once. Long
walks were taken. Plants and birds and fish, and the
manners and customs of the sparse inhabitants in turn
attracted observation, and kept the travellers busy.
Fishing was excellent everywhere. Fowling and shooting
were always to be had. Banks was ill with fever during
August, and had a milder attack in October ; otherwise,
he heartily enjoyed the whole affair. Even sea-sickness
was conquered in a series of boat trips, which were
frequent and necessary in the tiny archipelagoes that
fringed the coasts. Mosquitoes were very troublesome, a
source of trial to the temper which had not been foreseen,
and against which no precautions appear to have been laid.

The ship was stationed for some weeks in Croque
Harbour, and afterward visited Chatteaux Bay and other
settlements as far as Esquimaux Islands, off Labrador.
She returned southward in October, and sailed from
St. John's on the 28th of that month. The pleasure of
the homeward voyage was marred by a severe gale, during
which a sea broke over the quarter and almost filled
the cabin with water. A precious box of seeds was
demolished, as well as the box of earth with plants in it
which stood upon deck. Happily, the journal full of
zoological and botanical notes was kept intact.

The *Niger* made for Lisbon and reached that port on
November 2. She was stationed there for some weeks.
Thus Banks was enabled to stay in Portugal, and keep
his active mind at work on the many topics which ap-

pealed to him in that country.   He made fresh friends
there, and afterwards joined a club or society devoted
to the study of Natural History, with which he kept in
touch by correspondence after reaching home.   Early
in the spring his friend Captain Adams sailed for England.[1]
After his return, Banks divided his home into alternate
stay at Revesby Abbey and in New Burlington Street.
Upon his Lincolnshire estate he was arduous in studying
methods of agriculture, which, in his eyes, was really a
branch of botanical science.   He made his first appearance at the Royal Society on February 15, 1767.   The
Labrador expedition was thought well of in the scientific
world.   It was suggested to him that he would be expected to produce a report on the excursion for the
benefit of the Royal Society.   But nothing of the kind
was done : probably on account of shyness, a trait which
was visible in Banks at this period of his life and even
for some years later.

During these early years, Banks seems to have been
very fond of rambling about the country.   There was
always a tour in prospect as soon as summer impended.
It is much to be regretted that we know this only from
scanty inferences.   Two trips were undertaken this year.
Happily, there is record of one of these ;[2] a journey by
way of Dorset and Somerset to Bristol.   It is not clear
whether Banks had a companion.   His friend, Richard
Kaye, is mentioned once, and may have been fellow-
traveller all the way.   Banks was now adding archæology
to his means of culture.

He started May 15 for Dorsetshire, reaching Eastbury
House, in Tarrant Gunville, the same night.   There
was a barrow to be examined in the park.   At Chettle,

[1] There were journals of this excursion kept by Banks, which were
dispersed at the sale of his MSS. in 1886.   But a copy of the Newfoundland portion (in the handwriting of Miss Banks) is preserved at
the Natural History Museum.
[2] Printed by the *Bristol Naturalists' Society*, VI, pp. 6 *et seq.* (1899).

next day, there was another barrow, and on the 18th a third. He visited a distant relative, Mr. Bankes of Kingston Lacy, and saw his pictures ; and Mr. Humphrey Sturt, at Crichel, who kept several kinds of animals in the park. On the 21st there was another barrow. By this time he is prepared to launch out into conjectures as to their meaning and origin. Travelling through Shaftesbury and Bath to Bristol, he proceeded to Chepstow. Here, and in the neighbourhood, he revelled for several days, alternately charmed with the country and absorbed in botanizing. At Piercefield (" much improved since I saw it last ") he visited a friend, Valentine Morris. On the 26th he returned to Bristol, and had a day's botanizing at Clifton. After this he visited Wells, Glastonbury, Taunton, Bridgewater, Clevedon ; and then stayed at Clifton a few days, with more botanizing at St. Vincent's Rocks. On the way home to London he stopped to look at Silbury Hill and the other strange relics thereabouts, including the *Grey Wethers*, of which a few stones yet remain visible from the highway. He found " the people in that neighbourhood were breaking great numbers of them, either to mend the roads or build houses, which gave me an opportunity of examining them and bringing away some pieces, which I found to be of a very hard and fine-grained sandstone."

Another excursion was taken this summer, in the company of two companions. They spent a week or two in North Wales, mostly botanizing. William Hudson was one of this party ; in all likelihood a very good comrade for the occasion. His social merits were high, the outcome of a tranquil but genial disposition. He had studied with an apothecary, and practised medicine. An acquaintance with Benjamin Stillingfleet, together with an ardent study of the Sloane Collections, made him a botanist. He acquired a European fame by the publication of his *Flora Anglica* (London, 1762).

CHAPTER II

ROUND THE WORLD WITH CAPTAIN COOK

AN opportunity came, in the year 1768, for an exploit in all respects suitable to the active genius of Mr. Banks. Circumnavigation had been a topic of interest for several years past, not less in scientific than in political circles. British captains had gained renown, and Frenchmen had closely emulated them, by tales of adventure and research in hitherto unknown seas ; and bulky compilations of travel and exploration were popular with all classes of people who read any books at all. Now, yet another project was in the air, to unravel the mysteries of the southern ocean.

The occasion was this. The Royal Society was desirous of getting an accurate observation of the Transit of Venus, which was due in 1769. The Transit of 1761 had not been satisfactorily observed. Great anticipations were formed by astronomers of an improved result in 1769— Dr. Maskelyne at Greenwich, and the Rev. Thomas Hornsby at Oxford, led discussions on the more suitable stations for taking the observation. Hornsby contributed an excellent paper on the subject to the *Philosophical Transactions*. Having established their position, and marshalled its points in a memorial to the King, the Royal Society obtained an immediate acquiescence in the proposal to send various expeditions to distant quarters of the globe. On March 24, 1768, the President announced that £4000 had been paid to him, on account of the astronomical expenses. The places decided upon as

best for the observation were Madras, Hudson's Bay, and
an island in the Pacific Ocean. The Hudson's Bay Company granted a passage in one of their vessels ; and the
Admiralty fitted out ships for the use of the other parties.
The immortal James Cook, then a lieutenant, R.N.,
was chosen to take the company to the southern seas ;
accompanied by Charles Green, the astronomer. Alexander Dalrymple, hydrographer to the Admiralty, one of
the first geographers of the age, missed this chance of being
associated with this famous expedition : he was willing
to go in full charge, as it was first proposed ; but he insisted on the rank of captain, which the Admiralty refused, and his nomination was withdrawn.

Lord Sandwich was then at the head of the Admiralty ;
a man having every sympathy with the undertaking.
This circumstance had much to do with the generous and
ready aid of his Department. His young friend Banks
was not less concerned in the success of the affair. It was
not long before Banks came to think seriously of taking
a personal share in it. He proposed to the Council of the
Royal Society that he should join Lieutenant Cook ; and
it would appear, from their official application to the
Admiralty, that there was as much personal regard for
their patriotic Fellow as there was satisfaction with his
liberal offer.[1]

Banks's friends, generally speaking, were interested
and elated with the prospect of his voyage. But there
were some who urged him warmly to relinquish the
idea, and make the Grand Tour of Europe instead. His

[1] " The Council have appointed Mr. Charles Green, and Captain
Cook, who is commander of the vessel, to be their observers ; besides
whom, Joseph Banks, Esq., Fellow of the Society, a gentleman of large
fortune, who is well versed in Natural History, being desirous of undertaking the same voyage, the Council very earnestly request their Lordships, that in regard to Mr. Banks's great personal merit, and for the
advancement of useful knowledge, he also, together with his suite,
being seven persons more (that is, eight in all) together with their
baggage, be received on board of the ship in command of Captain
Cook."—Weld : *History of the Royal Society*, II, 38.

answer was, "Every blockhead does that; my Grand
Tour shall be one round the whole globe."

Several correspondents presented him with advice, not
to say elementary instruction. But there is a general
tone of real friendliness and appreciation. Mr. Pennant
warned him not to forget umbrellas (which were at that
time novel in England) and oilskin coats. Thomas
Falconer sent many pages of pure prolixity, on geography and geographers. A London merchant provided
him with peach-spirit; and with advice on the use of
lemon-juice and the proper way of curing fresh provisions
at sea. This gentleman had also a little bottle to offer
Banks, made from a "peculiar elastic glutinous body of
which small bottles and balls are made." Few persons,
in the year 1768, had seen caoutchouc; but Banks was
one of them, having brought from Lisbon two balls of it.
The Rev. Gilbert White had heard of the projected
voyage. He writes certain memoranda on Birds, and
wishes Mr. Banks a great deal of success and satisfaction
in his laudable pursuits; a prosperous voyage, and a safe
return. . . . "P.S.—I became somewhat of a botanist
without any teaching, and almost without books. But
under such a master as you are, I should be convinced
how little I knew."

Banks's preparations for his voyage were made on a
most ample, not to say extravagant, scale. No expense
was spared. The staff included John Reynolds, Sydney
Parkinson, and Alexander Buchan, artists; Henry
Sporing, assistant draughtsman; James Roberts and
Peter Briscoe, servants from Revesby; and two negro
servants. Beside these was Daniel Carl Solander, whose
acquaintance Banks had made in the preceding year.
His status was, perhaps, that of a friend and guest.
Solander was a Swedish naturalist who had come to
England on the suggestion of Linnæus. He made his
mark at once: was appointed a naturalist at the British

Museum, and became a F.R.S. in 1764. Few of the foreigners who then settled in England were so readily acceptable and so highly esteemed as Solander.

The *Endeavour*, Lieutenant James Cook, made sail from Plymouth on August 25, 1768, and anchored off Deal on July 12, 1771. During these three years she had sailed round the globe, in the following itinerary: Madeira, September 12 ; Rio Janeiro, November 13 ; Tierra del Fuego, January 15, 1769 ; Otaheite, April 13 ; New Zealand, October 8 ; New Holland, April 28, 1770 ; Torres Straits, August ; Savu, September 17 ; Batavia, October 9 ; Cape of Good Hope, March 12, 1771 ; St. Helena, May 1.

The results of this voyage were beyond expectation. The observation of the Transit of Venus ; many additions to Geographical knowledge ; the study of the products, and the fauna and flora of remote lands and seas ; and reports upon the character and condition of primitive peoples, were the definite objects of pursuit. In all these things the enterprise was entirely successful. This was due to the strong personality of each leader of the Expedition. Cook had proved himself a first-class marine surveyor in hitherto little-known quarters of the globe. Through his influence the crew were saved from the horrors of scurvy, and spared many of the ordinary hardships of a long sea-voyage. Banks was enthusiastic and untiring. His mind was active and vigorous, letting nothing escape his observation. Gifted with a manly presence and a genial but dignified manner, he usually impressed the untutored savage on very short acquaintance. By the recent publication of his own Journal of the Voyage,[1] light is for the first time thrown on his ability to treat facts synthetically, and to make a picturesque and intelligent summary of all that attracted his attention. For the first time, be it said : because, oddly

---

[1] Ed. by Sir Joseph D. Hooker (London, 1896).

c

enough, he seldom ventured into print for public reading
with the exception of a few horticultural or agricultural
papers.

Vast additions were revealed to European knowledge
of the plants, and the birds, and the fish, of tropical and
sub-tropical climes. Much of this accession of material
to Natural Science has been assimilated from time to
time ; only experts know very much about the details of
its source. Perhaps their greatest " wonder " was the
Kangooroo. The ship's company were interested to excite-
ment over this strange creature. Sometimes they made
it an object of pursuit when out in search of game.
Its movements upon the hind-legs, with a strange sort of
jump or hop, did not seem to lack speed ; for, on the
ship's greyhound being brought out to the chase, it could
not outstrip the Kangooroo.[1] Banks's manuscript
journal records faithfully the places whence he derived his
botanical treasures. The actual *Herbarium* reposes in
safety in the Natural History Museum at South Kensington.

A noticeable feature of this voyage was the general
good health maintained by the ship's company until they
reached Batavia. This was very much due to their
ability to get fresh and wholesome food whenever they
communicated with the shore. The captain was able
to give away his live stock when occasion offered, and

[1] A certain journal of high reputation, in reviewing Dr. Hooker's
Banks's Journal, remarked that " Dampier recorded, in 1697, how he
had eaten a sort of racoon with very short fore-legs, on the west coast of
Australia. It is a curious fact that the English, so ready to suspect
foreigners of taking advantage of their discoveries, are not always so
ready to give their neighbours all the credit due to them." This is both
unjust and in bad taste. The idea that the average Englishman ignores
the merits of the wise and clever men of other countries is a journalese
fallacy rather in vogue of late years. There has never been a people
so generous to the foreigner as the English. If this sarcastic allegation,
however, must needs be flourished once in a while, the very last person
to be chosen for exemplification should be Joseph Banks.

Besides, Dampier was no foreigner ! He was an Englishman, native
of Somerset.

many are the pigs and poultry which now flourish in Australasia and in the Friendly Islands that are descended from the stocks left behind by Cook. For months together the ship's provisions were kept almost untasted. Cocoa-nuts, and bread-fruit, and strange new birds, etc., supplied their larder. And, whatever length of time they were kept on or near the shore, there arose no distaste for these things. Perhaps the most remarkable of the new kinds of food which they enjoyed was the Bread-fruit (*Artocarpus incisa*), so plentiful in Otaheite that one could walk for miles under the grateful shade of these trees, which, together with cocoa-palms, were found growing everywhere in great profusion. They had scarcely landed at Otaheite, when barter began with the natives for bread-fruit, both roasted and raw. A bead as large as a pea purchased four or six bread-fruits and a like number of cocoa-nuts. Banks tells us that they were obliged to leave off buying for two days, so great was the supply. Two months later there was a sudden scarcity. The season had come for gathering the bread-fruits wholesale, with the object of storing them, or of manipulating them into a sort of paste—called *mahie*, which served for food until the appearance of a new crop. The chief sustenance of these people seemed to be the bread-fruit; and beside this, the inner bark of the tree furnished a capital material for weaving into cloth.

The ship's party were not exempt from occasional disaster. At Tierra del Fuego, the two negro servants succumbed to the cold and exposure; and here Mr. Buchan, the artist, exhibited the first signs of illness which carried him off three months later. Otherwise the entire company enjoyed good health until they stopped at the fatal port of Batavia, two years after leaving England. One after another they became ill with malarial fever. Several died, and were buried there; and others, after the ship left, who were buried at sea. These included

Green, the astronomer, who had successfully observed the Transit of Venus, and was thus deprived of the personal renown which awaited him at home. Monkhouse, the surgeon, had died on shore. Sporing and Parkinson followed. Cook was taken ill, and Banks and his friend Solander were brought very near to death's door. In all, thirty-eight persons perished during the voyage, most of them in consequence of the malarial climate of Batavia. In a social sense, the Expedition was fortunate everywhere, except at Rio Janeiro, where the authorities had been indisposed to entertain the voyagers, and even regarded their errand with suspicion. At the Dutch settlements in Java, people were extremely friendly and hospitable. And wherever they had to do with indigenous natives (which both Cook and Banks always call Indians) they lived on amicable terms. Any differences that did arise were owing to petty thefts. During their stay of three months at Otaheite there were no misunderstandings of a lasting character. One of the natives, Tupia, in company with his little son, joined the ship on leaving Otaheite ; but these unfortunately died at Batavia.

The ship herself escaped any serious trouble until she was among the reefs off the coast of Australia. On one day they had almost given up hopes of escape from utter shipwreck. What appeared to be a fatal leak, however, aroused the ingenuity of a midshipman, Jonathan Monkhouse, who suggested the operation of " fothering," i.e. preparing a sail in such manner as to cover the leak and arrest the flow of water into the ship's hold. On later examination of the damaged hull, it was found that a piece of rock had broken off and was fixed in the rent. This unique circumstance had really saved the ship, by delaying the influx of water in the first instance.

The *Endeavour* had been a long time absent from home. People began to wonder whether she would ever be heard

of again. Some messages had been received from the
ship's company early on the voyage. Mr. Pennant had a
letter from Rio Janeiro ; and then silence closed over
them. In October, 1770, there must have been unpleasant
rumour about them in the newspapers. Pennant, writing
to his friend the Rev. George Ashby, says, " I do not
know what to say about Mr. Banks. The account shocked
me greatly. What makes me uneasy is that I do not hear
anybody had a line from him from the Falkland Islands,
which were long in our possession after he touched at
them. I have wrote to his family, and hope to find the
newspapers contradicted."

All misgivings were set at rest in May, 1771, by the
arrival in London of news from Sydney Parkinson.
The poor man had written from Batavia, and his letter
reached England only two months before his more
fortunate fellow-voyagers. The *Endeavour* cast anchor
in the Downs on July 12. Mr. Banks landed the same
day and proceeded to London.

During these years of absence Banks was not forgotten
by his friends. Upon his reappearance there was universal
joy, especially " among the learned and curious." It
was not merely restoration to his own circle. Banks
leaped into fame, and became a person of importance.
The success of the hazardous voyage to which he had
committed himself, and the rumour of his achievements
in Natural History, caused him to be welcomed by all
that class of Englishmen who admire the results of hardy
and intrepid action. The Royal Society was not in
session, but the President, Sir John Pringle, speedily
carried him off to Kew, and introduced him to George III.
The King granted him an interview on August 10 ; when
Banks and Solander had a long conference with His
Majesty on the discoveries they had made, and their
marvellous adventures among the islands of the South.

A very cordial friendship arose now between Banks
and George III, a circumstance fraught with the greatest
benefits to the work of Science. His Majesty habitually
consulted Banks on points bearing on the welfare of
his people ; while it was the lot of Banks to have his
part in the institution of useful schemes, in a quarter
where there was likelihood of their being efficiently
promoted.

" The people most talked of at present are Messrs.
Banks and Solander."[1] He was certainly much lionized.
But the chief gratification was his reception by the
scientific world. Dr. Solander likewise came in for
proper recognition. Everybody liked Solander in his
personality alone ; to meet him at his official post or in
Banks's house was an intellectual treat.[2]

The friendship between Banks and Lord Sandwich was
largely consolidated by the great exploit. They were in
constant association, mutually interested in horticulture,
in useful public projects, and in the welfare of the Navy
especially. Sandwich was an able minister of State, and
made a splendid chief of the Admiralty.[3] Some time in

---

[1] Lady Mary Coke : *Journal*, III, 435 ; IV, 153, etc. This lady met
Banks in society several times. She throws a chance light on an unknown side of the young man's character : " August 11, 1771, Mr.
Morrice was excessively drole according to custom ; and said he hoped
Mr. Banks, who since his return has desired Miss Blosset will excuse his
marrying her, will pay her for the materials of all the work'd waistcoats
she made for him during the time he was sailing round the world."

[2] One of the distant admirers of Banks, who does not come closer
into his life, was the clever but wayward Edward Wortley Montagu the
younger. Writing to a friend (Feb., 1773) he says : " Good God ! how
happy are these gentlemen, in having been so serviceable to mankind.
. . . I am much obliged to you for the light in which you set me to Sir
John Pringle, Mr. Banks, and Dr. Solander ; but you diminish my
ardour to be acquainted with them, lest by knowing them they should
find me much below the high mark at which your friendship has placed
me. However, in the meantime assure them of the real gratitude with
which my heart is filled for their good opinion of me."—Nichols :
Literary Anecdotes, IV, 640, 645.

[3] This : in spite of the partisan traditions, which are reproduced
in one cyclopædia after another. Following Macaulay, writers in
allusion to Lord Sandwich roll under their tongues the delicious verbal

September of this year, there was company at Lord Sandwich's house at Hinchingbrook to meet Dr. Burney, Captain Cook, Mr. Banks, and Dr. Solander. Thus much Fanny Burney,[1] who has frequent notices of all these men. Few are the readers of our story but will regret there is no record come down to us of these delightful days : enlivened with the learning and anecdotage of Dr. Burney, the wit of Lord Sandwich, and the genial good-humour of Solander. On this occasion, too, the project of a second circumnavigation was discussed with young James Burney as a possible sharer in its perils and glories. It seems almost incredible that even the energetic Banks would care so soon again to meet the perils, and the enormous expense, of another voyage round the world, together with the enforced deprival of the society of his friends. Yet the thing was already in train. Before the close of the year, preparations were going forward for a second Expedition, on a more ambitious scale. At first it was proposed to send a fifty-gun ship and two frigates. But the decision was finally in favour of two vessels only, under the command of Cook.

Upon Lord Sandwich suggesting to Banks that his assistance would be welcome to the furtherance of the Expedition, he readily volunteered. His time was therefore occupied very busily during the ensuing months, partly in arranging his collections and treasures and exhibiting them to his friends, and partly in the prepara-

morsel, *Jemmy Twitcher* ; and thus perpetuate a loathsome and half-told story. We do not know all the truth about Lord Sandwich. His reputation has suffered from tales of his youthful follies ; all the more that exaggeration and caricature have made their memory endure. He was a steady friend. His social qualities were excellent, as we learn from Joseph Cradock and others. His dinner-table was a delight, as were his musical evenings. We might be tempted to give anecdotes in support of these statements ; but we must not be too discursive on matters apart from our central figure. Besides (since this note was written), it is reported that a careful and critical biography of Sandwich is already in hand.

[1] *Early Diary*, etc., I, 138.

tions for his new voyage. The same profuse expenditure as before was devoted to fresh stores of books, instruments, etc. A Royal Academician (Zoffany) was engaged as principal artist, together with three draughtsmen, two secretaries, and nine servants. Everything was done in a princely manner.

After all this preparation Banks was obliged to retire from personal share in the Expedition. There was so much difficulty about the accommodation to be afforded for his party, and the Navy Board being unable to meet Banks's views in several points, that he withdrew. Brougham and other writers who have alluded to this occurrence are rather warmly disposed to implicate individuals in the disagreements which occurred. It is not unnatural, under the circumstances; for the official mind seems to have been a little jealous of Banks's influence in the business.[1] Yet the difficulties would, perhaps, have been safely weathered but for an unforeseen circumstance, which precipitated matters.

Lieutenant Clerke, on board the *Resolution*, lying in the river, wrote to Mr. Banks (May 13, 1772) to the following effect : " We weigh'd anchor at Gravesend this morning about 10 o'clock, with a fine breeze from the eastward. The wind from that quarter laid us under the necessity of working down the reaches : which work, I am sorry to tell you, we found the *Resolution* very unequal to. . . . She is so very bad that the pilot declares he will not run the risk of his character so far as to take charge of her

---

[1] It is likely enough that more than one person already discerned in Banks a possible " despot." One matter in which he was thwarted was his nomination of Dr. Priestley. Banks invited him to join this second Expedition as astronomer. In his view an astronomer was an astronomer, and a great master in physics was a philosopher. But some clergyman on this Board of Longitude objected to Priestley's appointment on account of his religious principles ; and so strong became the opposition that Banks pressed the point in vain. (*v.* details of this incident in Kitson's *Captain James Cook the Circumnavigator*, London 1907.)

farther than the Nore without a fair wind ; that he cannot, with safety to himself, attempt working her to the Downs. Hope you know me too well to impute my giving this intelligence to any ridiculous apprehensions for myself. By God, I'll go to sea in a grog-tub, if required, or in the *Resolution* as soon as you please ; but must say I think her by far the most unsafe ship I ever saw or heard of. However, if you think proper to embark for the South Pole in a ship which a pilot will not undertake to carry down the river, all I can say is you shall be cheerfully attended as long as we can keep above water."

Between this 12th of May and the 31st, when Clerke wrote again from Sheerness, Banks had ordered all his stores to be removed from the ship ; Dr. Solander, Dr. Lind, Mr. Zoffany, and the draughtsmen did likewise. The unseaworthiness of the ship may have been occasioned by the alterations which had been made to accommodate the naturalists and their party. Be that as it may, a protest reached the Admiralty Office, in accord with the complaint made to Banks in this letter. In order to avoid the possibility of the King's anger at this denouncing of the ship, His Majesty was informed that it was all on account of Mr. Banks's unreasonable demands. Upon learning this, Banks instantly withdrew from the project.[1]

---

[1] Toward the close of his life, Banks was consulted by Robert Brown as to his recollection of the actual facts in this case. The following paper (in Brown's handwriting) is preserved, among his collected letters (vol. I), in the Natural History Museum. As the affair caused much discussion at the period, and it has always been left an incoherent story, it is worth while putting it here on record that Banks had a tale of his own to tell about it :—

" Soon after my return from my voyage round the world, I was solicited by Lord Sandwich, the first Lord of the Admiralty, to undertake another voyage of the same nature. His solicitation was couched in the following words : ' if you will go we send other ships.' So strong a solicitation, agreeing exactly with my own desires, was not to be neglected. I accordingly answered that I was ready and willing. The Navy Board was then ordered to provide two ships proper for the service. This they did and gave me notice when it was done. I immediately went on board the principal ship and found her very improper for our purpose.

How far Cook was involved in these disagreements is not known. It would appear that his relations with Banks were a little strained. He writes from the Cape of Good Hope (November 18, 1772) : " Dear sir, some cross circumstances which happened at the latter part of the equipment of the *Resolution* created, I have reason to think, a coolness between you and me. But I can by no means think it was sufficient to me to break off all correspondence with a man I am under many obligations to, . . ." and proceeds with a gossipy relation of some marine incidents, including a sad tale of some bad-

Instead of having provided a ship in which an extraordinary number of people might be accommodated, they had chosen one with a low and small cabin and remarkably low between decks. This I objected to, and was answered that it could not nor should not be remedied. With this answer I went immediately to Lord Sandwich, who having advised with several people ordered the cabin to be raised eight inches for our convenience, and a spar deck to be laid the whole length of the ship for the accommodation of the people. This order I suppose vext the Navy Board ; for, from that time they never ceased to pursue me with every obstacle they could throw in my way, and at last overthrew my design. First, to the proposed alterations they added a round-house for the Captain to be built over all this ; and all other alterations they made with timber so heavy and strong that the top of the round-house was literally thicker than the gun-deck of the ship. This though I saw I could not remedy. The ship was made so crank by it that she could not go to sea. Some of the oldest sea-officers, who I believe were jealous that discovery should go out of their line, procured an order that the ship might be reduced to her original state. In this situation then I was again offered the alternative to go or let it alone, with a great deal of coolness however ; for I now had inadvertently opened to them every idea of discovery which my last voyage had suggested to me, and thence they thought themselves able to follow without my assistance now they had once got possession of them.

" As the alterations which they had made rendered it impossible for my people to be lodged, or to do their respective duties, I resolved to refuse to go, and wrote a letter to Lord Sandwich, a copy of which is inserted in the Appendix, stating my reasons.

" I shall now give a list of the people who I had at my own expense engaged as assistants in this undertaking :—

" Dr. Solander: now well known in the learned world as my assistant in Natural History.

Mr. John Fredk. Miller
Mr. James Miller   } Draughtsmen for Natural History.
Mr. Cleveley

Mr. Walden   } Secretaries,
Mr. Bacstrom

pickled salmon, which even the pigs on board would not touch.

Banks still did what he could to promote the new Expedition. The naturalists were John Reinhold Forster and his son Georg. Both of these were really scientific men ; and later, after their return to Germany, acquired deserved fame for their philosophical writings. The father had come to England in 1766, with a young family, and had a rough struggle for existence, until he came under the patronage of Daines Barrington, Thomas Pennant, and other votaries of Natural Science.

One inevitable result of the exploit of Captain Cook was the appearance in the air of the literary birds of prey. There was money to be made out of the Tale. In the magazines and elsewhere there were feeble attempts made to anticipate the authentic narrative, a circumstance which caused some annoyance to Mr. Banks and his friends.

besides nine servants, all practised and taught by myself to collect and preserve such objects of Natural History as might occur ; three of whom had already been with me on my last voyage. Besides this I had had influence enough to prevail with the Board of Longitude to send with us Messrs. Bailey and Wales as astronomers, and also with the House of Commons to give £4000 to enable Dr. Lind of Gorgie, remarkable for his knowledge in Natural Philosophy and Mechanicks, to accompany us.

" These gentlemen, except only the astronomers who did not at all belong to me, were to a man so well convinced of the impossibility of our going out in the state the ship was now reduced to, that they all refused with me, and so well were they satisfied with my conduct that though I believe every one but Dr. Solander was separately tampered with to embark without me, not one would at all listen to any proposals which could be offered to them.

" Upon my refusal to go out, the ships were ordered to proceed, and in order to do as much as possible even in the branch of Natural History, Mr. Forster, a gentleman known to the learned world by his translations of several books, was engaged under the immediate protection of the King ; and, soon after, Mr. William Hodges, a young man who had chiefly studied architecture, was joined to him as Landscape and Figure Painter. With those gentlemen on board, the ships *Resolution* and *Adventure* sailed from Plymouth on the 12th of July, 1772. In the meantime I had received several overtures from the East India Company, who seemed inclined to send me on the same kind of voyage the next spring."

One of these publications appeared as early as September, 1771 : *Journal of a Voyage round the World in the Endeavour*.[1] This was published by Thomas Beckett, bookseller in the Strand, who in capacity of Editor, says, " I considered to [publish] it from the agreeable manner in which it was written, as well as by the honourable mention that is made of the ingenious gentlemen, Mr. Banks and Mr. Solander, and I am convinced it is the production of a gentleman and a scholar." The individual's name was never divulged. No less than ten private journals were handed over to Cook at his request, but this " gentleman " appears to have evaded it. The special interest attaching to the book occurs in the narrator's allusion to the disagreeable consequences resulting from the behaviour of some of the crew, while on shore at Otaheite ; and the inference drawn that a French crew had been there before them.

Perhaps the writer of this journal (if not the person who put it into literary form) is B. Lauragais, who wrote to Banks in the spring of 1772. Alluding to the reproaches he was expecting, he says, " which, I think, I am so little deserving of, that when I informed you lately of the consent I had given to the printing of the French *galanteries* at Otaheite, I imagined you answered you would be glad of it : considering that nothing was contained in that little narrative but what was true, and at the same time honourable to you. . . . I believe that less disadvantage can arise from printing our writings than in permitting Thousand Falsehoods to be spread abroad supposed to be related by us. Being actuated by these motives, I have permitted the printing of them ; and my consent was solicited by a man to whom I had given my manuscript. . . . The printer is so eager to sell them that he does little care for correction."

[1] Transl. into French by A. F. J. de Fréville : *Supplément au Voyage de Bougainville*, etc. (Paris, 1772).

Another account of the Expedition appeared the same year : "A Journal of a Voyage to the South Seas in H.M.S. the *Endeavour*. Faithfully transcribed from the papers of the late Sydney Parkinson, draughtsman to Joseph Banks, Esq., on his late Expedition with Dr. Solander, round the world. Embellished with views and designs delineated by the Author, and engraved by capital Artists."

This book gave Banks a great deal of trouble. Parkinson was quite a young man, with considerable promise as an artist. Banks had been generous toward him, always encouraged him, paid him well, and was much grieved at his untimely death. But he supposed that the result of Sydney's work was his own property, as having been done in his service. Imagine his surprise, on learning that a brother, Stanfield Parkinson, laid claim to Sydney's drawings and collections ! When the above work was published, readers were indulged to a lengthy preface, full of the Editor's wrongs with Mr. Banks ; with an acrimonious account of their dispute ; and plainly charging Banks with embezzling his brother's property : albeit throwing in some admissions of his " generosity, integrity, and probity." The dispute widened and lengthened. Dr. Fothergill (hitherto unacquainted personally with Banks) was called in as a sort of arbitrator. After a weary time, during which Banks maintained the rights he had claimed, and Parkinson quarrelled with Dr. Fothergill, the affair came to an end, and Banks gave £500 to Sydney Parkinson's family. Stanfield was evidently rather a crazy person. He presently died in St. Luke's Hospital.

Of the ten journals which were delivered up to Captain Cook, seven have been printed, recently, in the *Historical Records of New South Wales*, vol. I. One of these was by Green, the astronomer ; another by Charles Clerke, who was concerned in the later voyages of Cook.

It was Banks's intention to produce on his own account a history of the Expedition and its results in Natural Science. He proceeded very far with this undertaking. A vast number of drawings and paintings were finished off from the rough sketches of Buchan and Parkinson ; and people were asking, for years afterwards, especially from Germany, when the publication would be complete. The reason why it was relinquished has never been made clear. The drawings, or very many of them, are in the custody of the British Museum. There are five hundred finished copper plates stowed away at South Kensington. A partial resurrection has been made, in a superb volume, " Illustrations of Australian Plants, collected in 1770 during Captain Cook's voyage round the world in the *Endeavour*. By the Right Hon. Sir Joseph Banks, Bart., K.B., P.R.S., and Dr. Solander, F.R.S. With determinations by James Britten, F.L.S., Senior Assistant, British Museum. Printed by order of the Trustees . . ." (Folio, London, 1905).

What may be called the official account of Cook's first voyage was undertaken by John Hawkesworth, who received from the Government £6000 in payment for his services. His story was made up from Cook's journal, Banks's own journal, and some personal contributions from Dr. Solander, assisted by access to Admiralty records. The book was one of the literary triumphs of the day. It has naturally become classic to the geographer. But it was fatal to Dr. Hawkesworth. Envy and jealousy marked him. A fierce attack was made upon him ; the principal charges being that he had allowed himself an occasional exhibition of Free-thought, as when he omitted to regard a fortunate escape from peril as a special interposition of Providence ; and that he had shown an absence of reserve, in dealing with matters which the Grundys of the day thought should have been marked with a specially thick veil. These things would

JOSEPH BANKS, ESQ.
From a mezzotint by Dickinson, after Sir J. Reynolds. Circa 1774

now pass without notice, unless as the stalking-horse for a little humour. Indeed, it would be difficult to find in Hawkesworth's volumes the sources of complaint, unless aided by taking down from dusty, musty shelves the long-forgotten gibes. But they amounted to Disgrace, with a man of finely sensitive temperament. He died broken-hearted in November, 1773.

Cook's own journal has been edited in our own time, in a manner suitable to an age which demands a scientific performance of such work : by Captain J. L. Wharton, R.N., Hydrographer to the Admiralty (London, 1893).

CHAPTER III

VISIT TO ICELAND. COOK'S SECOND AND THIRD VOYAGES

THE preparations which Banks had made for a second voyage to the South Seas were not destined to be entirely in vain.

A Swedish clergyman was in England on a travelling tour, Uno von Troil by name. He was interested in Scandinavian antiquities, and later came into some eminence as Archbishop of Upsala. Having made acquaintance with Banks, it was proposed that they visit Iceland in company. The project was immediately adopted. Measures were taken for its organization in that profuse style in which Banks thought proper to undertake a scientific excursion. A ship was chartered, at a cost of £100 per month, and set sail on July 12, 1772, with a company of forty persons. Banks's guests included the inseparable Solander; Dr. von Troil; James Lind, a rising young physician of Edinburgh, who was, besides, an astronomer; J. F. Miller, artist and engraver; and Lieutenant Gore, who had been with him in the *Endeavour*.

The party landed for two days in the Isle of Wight (which von Troil calls a little paradise), proceeded to Plymouth, and then made for the Western Islands of Scotland. They were lying in the Sound of Mull early one morning, near the seat of a Mr. Maclean, of Drumnen, who, upon the accidental acquaintance, invited Mr. Banks and his friends to breakfast. There was another English gentleman, a Mr. Leach, present at the table, who mentioned in course of conversation that there was an island out in the open, nine leagues away, which he believed had scarcely ever been visited. The long-boat was forthwith prepared, and a small tent and two days' provisions were placed on board. The ship remained near Tobermory, while her boat made its way to Staffa with Banks and his friends. Thus was Staffa discovered, and became a notoriety. Banks gáve a very full account of the island, including elaborate measurements, in his journal: which von Troil made use of later, and which Mr. Pennant copiously adopted in his *Journey to the Highlands*.

The party reached Bessestedr, in Iceland, on August 28, and remained on shore for about a month. The itinerary was (1) from Havnefiord to Heder Bay, (2) Laugarvatn, (3) Mola, (4) Mola, (5) Skalholt, (6) Skard, (7) Graufell, (8) Skard, (9) Straungiörde, (10) Reikjavik, (11) Havnefiord. Thus their tour embraced many of the most remarkable features : Thingvalla gaa, the Geysers, Mount Hekla, the Hvitaa, etc.[1] Mr. Banks impressed the islanders deeply with his personality. In after years they had further reasons to be mindful of his high and generous character. Solander and von Troil came in likewise for a great deal of regard. Perhaps the liberal and almost princely style of their travel had something to do with this ; but it was the men of learning in the island who displayed the most gratification with these visitors. Several odes were composed in honour of Banks and his friends. A great number of plants and other objects in natural science were collected and brought home. Beside the botanical treasures, their spoil included some Icelandic manuscripts which had been purchased by Banks.

Sir Joseph is one of the few travellers [? the first] who have left record of the ascent of Mount Hekla :—

" We ascended Mount Hekla with the wind blowing against us so violently that we could with difficulty

[1] Solander MSS., N.H. Museum.

D

proceed. The frost, too, was lying upon the ground, and the cold extremely severe. We were covered with ice in such a manner that our clothes resembled buckram. On reaching the summit of the first peak, we here and there remarked places where the snow had been melted, and a little heat was arising from them ; and it was by one of these that we rested to observe the barometer, which was 24°.838. Thermometer 27°. The water we had with us was all frozen. Dr. Lind filled his wind machine with warm water ; it rose to 1°.6, and then froze into spiculæ, so that we could not make observations any longer. We thought we had arrived at the highest peak ; but soon saw one above us, to which we hastened. Dr. Solander remained with an Icelander in the intermediate valley ; the rest of us continued our route to the summit of the peak, which we found intensely cold ; but on the highest point was a spot of three yards in breadth, whence there proceeded so much heat and steam that we could not bear to sit down upon it."[1]

The journey, homeward, was taken leisurely. The party wandered some time among the Highlands and Islands of Scotland, and paid many visits. It was not until November 19 that Banks left Edinburgh for London, accompanied by Dr. Solander and Dr. Lind.

A party of " Esquimaux Indians " was in London in 1772-3. Banks, of course, went to meet them ; the rather that he had missed seeing any when he was in Labrador. The usual enthusiasm of Londoners for a

---

[1] MS. Journal, quoted in W. J. Hooker, II, 116. It is very much to be regretted that Banks's Journal of this time is missing, or lost. When von Troil published *Resa till Island* [Upsala, 1777. Transl. by J. R. Forster as *Letters on Iceland* : London, 1780] it was found not to be an itinerary, but a series of letters or essays, describing the country and some of the incidents of travel, but with scarce a date or locality mentioned. There is a noticeable lack of allusion to Mr. Banks. One welcome memorial of this trip, in Miller's drawings of plants, together with coloured developments of his sketches by Thomas Burgis (1776), is preserved in the Natural History Museum.

new show was displayed, and the Indians were exceedingly popular. They had lodgings in Leicester Street ; and were taken to see St. Paul's Cathedral, and the Tower (as Cookists of the period). But they could not be made to understand much ; London Bridge, for example, they looked upon as an abnormal piece of rock. They pined for their glaciers and their snow-fields. At length, on the eve of sailing away from Plymouth, the two men died from fever and the two wives and a child went home.

This episode brings into our pages a very interesting and adventurous character : George Cartwright, author of *Journal of Transactions during a residence of sixteen years in Labrador.* Cartwright came from Nottinghamshire. He was a gentleman cadet at Woolwich, one of that class which does not learn its lessons, and is consequently handicapped for life. He was in the East Indies for a time, and later at the wars in Germany. Poor, and evidently unthrifty, but fond of shooting and sport, he drifted to Newfoundland, apparently for the simple reason that his brother John was going thither, as commander of the *Guernsey,* fifty guns. This was in 1765. He must have met Banks and Phipps during their stay. He made six voyages out and home during his sixteen years. It was on the second trip that he brought the Indians to England. Cartwright appears to have been under some money obligations to Banks, for there is an expression of thanks in one of his letters for " giving him time to discharge his bond." He sent a few curiosities home to Banks, but speaks of them with apathy. He never made a fortune, yet he doubtless enjoyed life in his own way. A Hudson's Bay Company's post, south from Hamilton Inlet, bears Cartwright's name to this day. A portrait of him, in the costume of a trapper, adorns the first volume of his *Journal.*

Cartwright's short letters to Banks are sufficiently doleful. But the writing of his book seems to have made him happy, particularly when Banks "expected both entertainment and instruction " from the volume.

Banks went beyond the mere curiosity-hunters, as was the way with him. The relations of the Eskimos with other races formed the subject of erudite discussion with his friends. Among his correspondents were Dr. James Hutton, the geologist, a man who then was leading the van in ethnographical inquiry ; and Dr. William Robertson, the historian, who requests Banks to entertain him with all the curious observations he has made upon men in an early stage of culture ; and introduces to his notice the novel speculations of Lord Monboddo concerning the Simian ancestry of mankind. Monboddo maintained that there were, in all likelihood, existing races of men with rudimentary tails. Banks appears to have treated his lucubrations with some respect, and corresponded freely in an open-minded way.[1]

In March, 1773, Banks was on a visit to Rotterdam, in company with the Hon. Charles Greville. They attended a meeting of the Batavian Society. Banks told them of his wish to undertake a voyage toward the North Pole, and desired that they would communicate with him on the discoveries and observations that the Dutch nation had already made, as far as 84° N. He promised in return to furnish them with ample information concerning the results of his trip.

This Dutch excursion was the means of adding several good acquaintances to his circle.[2] The visitors made a profound impression upon their new friends. One of them wrote afterward to Banks with some warmth : " Your visit leaves a kind of sensation hard to describe.

[1] *Hist. MSS. Com.*, VI, pp. 674, 678.
[2] A Journal of this jaunt was among the Banks manuscripts dispersed at Sotheby's sale in 1886.

It is like the remembrance of a pleasing dream ; with this difference, however, that it will not be transient." There was a trip to Wales in the ensuing summer. Banks went in company with Solander and Dr. Blagden. Lightfoot joined them at Chester ; a most industrious botanist was he, wandering many miles about England in pursuit of his favourite study. He had been with Pennant on his tour to Scotland in 1772, who made him " happy in his company, and himself agreeable to the several families who honoured me with their hospitality."

The fourth in this party was Charles Blagden, a young Edinburgh physician, the junior of Banks by four or five years, now at the beginning of a career as experimental philosopher which brought him to some distinction.[1] Blagden presently became closely intimate with Banks, and was one of his most assiduous correspondents during periods of absence. When at Revesby in the summer Banks was thus kept in touch with what was going on in town. Early in 1776 Blagden was appointed physician to the forces sent to America. He wrote to Banks frequently, sending abundance of botanical and ornithological notes, and observations on Natural Science ; besides shrewd remarks on the progress of the war, and the hospital concerns that filled his professional time. The length of his letters was prodigious, sometimes extending to eight or ten folio pages, about birds and quadrupeds and fishes. He sent large collections in Natural History to England, some for Daines Barrington and some for Banks. The former handed over these

[1] " Talking of Dr. Blagden's copiousness and precision of communication, Dr. Johnson said, ' Blagden, Sir, is a delightful fellow.' " —Croker's *Boswell* (1860), p. 663.
This painstaking editor adds, from Hannah More's *Memoirs*, II, 98 : " Doctor Blagden is so modest and so knowing that he exemplifies Pope's line, ' Willing to teach, and yet too proud to know ' " (which is a parody, of course, on " Willing to wound, and yet afraid to strike ").

treasures to Sir Ashton Lever, after giving Banks the opportunity of selecting from them.

Banks was elected into the Dilettanti Society in 1774, and was Secretary from 1778 to 1797, a post he held with "much satisfaction and advantage." The marbles of the Society were kept in his custody at Soho Square.

After the death of the Founders, the prominent members of the Society were Banks, Greville, Sir Wm. Hamilton, Sir R. Worsley, Sir George Beaumont, Sir Henry Englefield, Charles Townley, R. Payne Knight. (Cust: *History of the Dilettanti Society*.) There are two portrait groups still in the possession of the Society, painted by Sir J. Reynolds. One of these is reproduced on the adjoining page.

A new friend of 1774 was the Rev. Sir John Cullum, of Hardwick Hall, Suffolk, and Rector of Hawsted, of which parish he wrote an excellent history. Sir John was one of the many country parsons of the period against whom some discredit has been aroused because they were not "enthusiasts" (as the term was); but pursued the honourable though lowly task of minding their flocks, while not neglecting the mental culture without which life was hardly worth living. The country was full of good and hearty performance of duty, in a quiet way: that way, indeed, which does not make notorious History, but toward which we are beginning to take a more fervent regard, as of simple annals which really concern the life of man more than all the fussy acts of politicians and soldiers. We should be the better and wiser for a more intimate knowledge of some of these humbler careers, the records of which lie buried in many a country house or parsonage.

Sir John Cullum was becoming a good Botanist. He certainly was enthusiastic over that. He rejoices to find that Lord Herbert of Cherbury reckoned "Botanique" among the qualifications of a gentleman; "it gives one a pleasure to find other persons, of whom we entertain a good opinion, engaged in the same pursuits as oneself."

From meeting Banks and Solander at the Museum, a cordial acquaintance sprang up between them and Cullum. Botanical letters went to and fro, and Sir John was a devoted attendant in Banks's library. His circle of botanical friends widened. Among others, Mr. Lightfoot met him, and became a frequent correspondent and co-labourer in Science. Sir John Cullum presently revealed to Banks his wish to join the Royal Society. As the latter regarded him as an example of the man who deserved such recognition of original research, Sir John's nomination was promptly made.

Another new Fellow of the Royal Society about the same time was Dr. Alexander Hunter, of York, who was then preparing the fine edition of Evelyn's *Silva*, with which his name has since been associated. It is likely enough that this is one of the cases where Banks gave financial assistance. Dr. Hunter's frequent consultations and evident reliance on the judgment of Sir Joseph justify this conclusion.

James Bruce, the African traveller, was a new acquaintance early in 1774. A similarity of pursuits and inclinations drew him and Banks together, and there was some occasional correspondence between them which seems to betoken very friendly relations.

The second expedition of Captain Cook had not been heard of since Cook wrote from the Cape in 1772. Without any actual fear existing as to the fate of the great seaman after two years' absence, people were beginning to wish earnestly for some news of the Expedition. The veil was partially lifted on July 14, 1774, when one of the ships arrived at Spithead.

The squadron, which consisted of the *Resolution*, Captain Cook, and the *Adventure*, Captain Tobias

Furneaux, had left Plymouth on July 13, 1772. The objects they set out for were duly accomplished without any serious disaster, excepting the loss of an *Adventure's* boat, which fell into the hands of cannibals. Cook had made a point of ascertaining if land existed in the far South. As far as 71° S. the question was settled in the negative. Visits were paid to New Zealand, Otaheite, and other South Sea Islands, where they encountered many old friends, and a few hostile " Indians." New surveys were made, and a large accumulation was made in furtherance of Natural Science. Forster wrote once to Daines Barrington, to whom he reported 260 new plants and animals. Others of the scientific side of the Expedition were, Wales, the astronomer and mathematician ; William Bayley, astronomer on the *Adventure;* William Anderson, the surgeon's mate ; who made their mark in life. George Vancouver, a middy on the *Resolution,* and James Burney, lieutenant of the *Adventure,* were also heard of again with honour.[1]

More than once the *Adventure* was separated from her consort. In the end Captain Furneaux thought proper to make for home alone, and so reached England a year in advance of the *Resolution.* Her arrival provided an

---

[1] When the *Resolution* was at Otaheite, there were two concurrent plots for staying there altogether and letting Cook get home how he might. This story (hitherto unrecorded) is told by one of those concerned. The writer has just heard of the mutiny of the *Bounty:* " Something what Bligh's people did was designed by most of the people of the *Endeavour* [this is a slip of memory] headed by Anderson and Gray, I think. They were for remaining, which prevented two or three gentlemen from doing so. When the scheme was discovered, the only successful argument against it was getting a certain disease and dying rotten . . . it turned the scale ; otherwise Cook, with the two superior messes must have found his way home had the ship been spared. I was a ringleader among a few who had prepared for remaining." (J. M. Matra to Sir Joseph Banks, May 7, 1790. Addl. MSS., 33979/29.)

This Matra must be identified with Marr, or Marra, an Irishman (who seems to have passed for a Dane) who joined the *Endeavour* at Batavia in 1770, and went in the *Resolution* on Cook's second voyage. (*v.* A. Kitson : *Capt. James Cook,* p. 205.) We shall hear again of this enterprising fellow.

entirely new sensation, in the shape of a real live Otaheitan.
Furneaux had taken on board a young native chief, by
name Omai. When Cook reached England in the following
summer, he found Omai a darling of Society. Under the
auspices of Lord Sandwich and Mr. Banks, the untutored
one was received everywhere with gratification. It must
be said that Omai justified his entrance into the world
of Lords and Ladies by his natural dignity of manner,
and by his adaptability to some of the inevitable veneer-
ing. Polish cannot be acquired by the stupid ; and
Omai was far from that. As Miss Burney says, " his
manners are so extremely graceful, and he is so polite,
attentive, and easy, that you would have thought he
came from some foreign Court."[1] Dr. Johnson was also
struck by his good manner, on meeting him at the
Streatham dinner-table.

This forgotten episode is often the subject of gossip
and pleasantry in the chronicles of the time. Fanny
Burney is very entertaining on the topic. The following
(unpublished) memorandum of Sir John Cullum is clear
and precise, and as good as anything that was ever printed
of Omai. And it is worth introducing here as an example
of the baronet's charming style. " (3 Dec., 1774). I
have had opportunity this week of being twice in company
with Omai, a native of Waietea, brought into England
this year ; of whom, from observation and enquiry, I
collected the following notices. He is about 30 years old,
rather tall and slender, with a genteel make ; his nose is
somewhat flat, and his lips thick, but on the whole
his face is not disagreeable. His ears are bored with a
large hole at the tip ; his complexion swarthy ; his hair
of considerable length, and perfectly black. The backs
of his hands are tattowed with transverse lines, and his
fingers with round ones ; the lines are not continuous,
but consist of distinct bluish spots ; his posteriors are

[1] *Early Diary of Frances Burney*, I, 334 *et seq.*

the only other parts tattowed. He walks erect, and has acquired a tolerably genteel Bow, and other expressions of civility. He appears to have good natural parts ; has learned a little English, and is in general desirous of improvement. Particularly he wishes to learn to write, which he says would on his return enable him to be of the greatest benefit to his country. But I do not find that any steps have been taken toward giving him any useful knowledge ; Mr. Banks seeming to keep him as an object of curiosity, to observe the workings of an untutored, unenlightened mind. When he is serious, and observing what others are saying, his Look is sharp and sensible, but his Laugh is rather childish. When he wants you to understand something he has seen, he uses very lively and significant gestures ; and is in truth a most excellent Pantomime. He is pleased (as many of more improved understandings often are) with trifling amusements, and is unhappy when he has nothing to entertain him. When I dined with him, with the Royal Society, a small magnifying glass had been newly put into his hands ; he was perpetually pulling it out of his pocket, and looking at the Candles etc. with excessive delight and admiration. We all laughed at his simplicity, and yet probably the wisest person present would have wondered as much, if that knick-knack had then for the first time been presented t him. He had seen Hail before he came into England, and therefore was not much surprized at the first fall of Snow, which he called, naturally enough, white Rain. But he was prodigiously struck, when he first saw and handled a piece of Ice ; and when he was told that it was sometimes thick and strong enough to bear men, and other great weights, he could scarcely be made to believe it. He is entirely reconciled to the European manners and customs. He conforms to our diet, which he likes very well ; and denies (against self-conviction) that his countrymen eat human flesh. He drinks wine,

OMAI, THE OTAHEITAN
From a mezzotint by Jacobi, after Sir J. Reynolds

but is not at all greedy of it ; and has never been intoxi-
cated since he was in England. He likes the English
women, particularly those of a ruddy complexion, that
are not fat. He submits most readily to the slightest
controul, and has not the least appearance of a fierce and
savage temper. I observed him play with a gentleman
who sat by him, and encouraged him, with all the cheerful
and unsuspecting good-nature of childhood. The King,
with much humanity, ordered him to be inoculated for the
small-pox, last summer, at his own expense. The fine
print of him, engraven by Bartolozzi, from a drawing by
Dance, is extremely like him."

The Rev. Michael Tyson,[1] at Cambridge, saw a good
deal of Omai. He says : " There was an openness of
countenance, and a native politeness, that would do
honour to an Englishman. . . . The Bishop of Lincoln
[John Green] was much in his company, and he found the
two leading principles of his mind were a regard for
Religion and a desire for Revenge. He was particularly
offended at the Bishop sitting at table between two
ladies : a custom not allowed the High Priests in his
country ! "

As Banks was absent with a yachting party—which
perhaps included Captain Phipps, Omai, Miss Ray,[2]

[1] Afterward Rector of Lambourn, Essex ; another clergyman of the
same type. He was a botanist and a scholar, beside being a good
parish priest. Sir John Cullum said he was far the best correspondent
he ever had : and they were numerous.

[2] This is Martha Ray, whose unlicensed union with Lord Sandwich
was tolerated by his friends because of her amiable character and her
charming behaviour in society. Some men ventured to insult her
covertly, but they were few in number. A bishop's lady, present at one
of their parties, avowed her regret that she had sat opposite to Miss Ray,
but found it " improper " to notice her : " she was so assiduous to
please, was so very excellent, yet so unassuming, I was quite charmed
with her, yet a seeming cruelty to her took off the pleasure of my
evening " (Joseph Cradock, *Memoirs*, IV, 168). Cradock himself writes
always in her praise. Banks and his friends were happy enough in her
company. Lord Sandwich was a better man for the association.
  Miss Ray was murdered at the theatre one evening, by a rejected
admirer, James Hackman. The culprit appealed to Lord Sandwich for

and for part of the time, Lord Sandwich—he was not at hand with his congratulations when the *Resolution* came into port.

Daniel Solander to Joseph Banks.

" London, August, 1775. . . . This moment Cook is arrived. I have not yet had an opportunity of conversing with him, as he is in the Board-room. Give my compliments to Miss Ray, and tell her I have made a visitation to the birds, and found them well.

" . . . Captain Cook desires his best compliments to you. He said nothing could have added to the satisfaction he has had in making the Tour, but having had your company."

" London, August 14, 1775. My dear Sir, an expedition down to the *Resolution* made yesterday quite a feast to all concerned. We set out early from the Tower, reviewed some of the Transports, visited Deptford Yard, went on board the *Experiment ;* afterwards to Woolwich, where we took on board Miss Ray and company, and then proceeded to the gallions ; where we were welcomed on board the *Resolution,* and Lord Sandwich made many of them quite happy. . . . All our friends look as well as if they had been all the while in clover. All enquired after you. In fact, we had a glorious day, and longed for nothing but you and Mr. Omai.

" Lord Sandwich asked the officers afterwards to dine with us at Woolwich. I had not much time to see the curious collections. Mr. Anderson, one of the surgeon's mates, has made a good botanical collection. On board are three live Otaheite dogs, ugly and stupid ; a spring-

pardon, and not in vain. Sandwich sent word to him that "as he looked upon his action as an act of frenzy he forgave it ; that he regarded the stroke as coming from Providence, which he ought to submit to ; but that he had robb'd him of all comfort in the world."— Delany : *Autob. and Corresp.,* IV, 424.

bok from the Cape ; a Surikate, two Eagles, and several small birds. My best compliments to Captain Phipps, August, and Omai."

All this looks as if Mr. Banks was, at present, keeping aloof from the circumnavigators. There was a dinner at Greenwich with Dr. Maskelyne, on September 5. The company included Solander, Captain Cook, Sir John Pringle, General Roy, and others. (" Company of Banks and Captain Phipps much wished for.")

Daniel Solander to Joseph Banks.

" London, August 21, 1775. My dear Sir, Mr. Harlock has sent to your house the plants I mentioned in my last letter. They are collected near Tranquebar by the Brethren of the Moravians, and as good specimens as I have seen. . . . Several of the *Resolution's* men have called at your house, to offer you their curiosities. Tyrrell was here this morning. . . . Captain Cook has sent all his curiosities to my apartments at the Museum. All the shells are to go to Lord Bristol. Four casks have your name on them, and I understand they contain Birds and Fish, etc."

John Marr to Joseph Banks.

" from on board H.M.S. *Resolution*, 1775.

" Begging pardon for my Boldness. I take this opportunity for acquainting your Honour of our arrival. After a long and tedious Voyage. Having met with extraordinary good success to the S'd and elsewhere, from many strange Isles I have procured your Honour a few curiosities as good as could be expected from a person of my capacity. Together with a small assortment of shells. Such as was esteem'd by pretended Judges of Shells. We have many experimental men in our ship that pretend to know. What was never known. Nor yet never will be known. I have something extraordinary to relate to your

Honour. But. A good opportunity will soon offer I hope. Depend upon it, Sir, I shall Take special care of Sending the above mention'd articles. When in order and an opportunity serves.

" Interim I remain your very Hble. Servt."

The yachting excursion above mentioned was not the only one taken this summer. Banks formed one of a party to a jaunt in Yorkshire among the Cleveland hills ; the story of which is narrated by that fine anecdotist George Colman the younger. The travellers were Captain Phipps ; his youngest brother Augustus, a boy of Colman's own age (i.e. about thirteen) ; the elder Mr. Colman and his son ; Mr. Banks, and Omai, of the Friendly Isles. The three seniors were intimate friends.

They started from York in Banks's carriage, " as huge and heavy as a broad-wheeled wagon," though but just large enough to contain the six passengers and their abundant luggage. The books of Captain Phipps, who was afterwards to make a stay at Mulgrave, and his boxes and cases crammed with nautical lore ; maps, charts, quadrants, telescopes, were put in, like stores for a long voyage. Banks's stowage was still more formidable. He travelled with trunks containing voluminous specimens of his *hortus siccus* in whity-brown paper, and large receptacles for further accumulation of vegetable materials. The vehicle had other characteristic encumbrances. One of these was " a remarkably heavy safety-chain—a drag-chain upon a newly-constructed principle, to obviate the possibility of danger in going down a hill : which snapped short, however, in our very first descent, whereby the carriage ran over the post-boy who drove the wheelers, and the chain of safety nearly crushed him to death." There was, besides, a " hippopedometer," by which a traveller might ascertain the precise rate at which he was going, in the moment of his consulting it.

" This also broke in the first ten miles of our journey, whereat the Philosopher to whom it belonged was the only person who lost his Philosophy." Their progress was much retarded by Banks's indefatigable botanizing : " We never saw a tree with an unusual branch, or a strange weed, or anything singular in the vegetable world, but a halt was immediately ordered : out jumped Mr. Banks, out jumped the boys and the Otaheitan . . . many articles, all a-blowing, all a-growing, which seemed to me no better than thistles, and which would not have sold for a farthing in Covent Garden Market, were pulled up by the roots, and stowed carefully in the coach as rareties."

Arriving at Scarborough, young Colman saw the sea for the first time. He was making his maiden plunge from a bathing-machine, when Omai appeared before him, wading. " The early sunbeams shot their lustre upon the tawny Otaheitan and heightened the cutaneous gloss which he had already received from the water. He looked like a specimen of pale moving mahogany highly varnished—tattowed with striped arches, brown and black, according to the fashion of his country. He hailed me with the salutation of *Tosh !* which was his pronunciation of George,"[1] and proposed to take a swim with George on his back. The boy consented, and was delighted with the result. " I made my way as smoothly as Arion upon his Dolphin. I could not indeed touch the Lyre, nor had I any musical instrument to play upon, unless it were the comb which Omai carried in one hand, and which he used while swimming to adjust his harsh black locks hanging in profusion over his shoulders. . . . My father looked a little grave at my having been so venturous. The noble Captain and the Philosopher laughed heartily, and called me a tough little fellow ; and Omai and I henceforth were constant companions."

[1] Omai's address to the King, on being introduced by Lord Sandwich, was " How do King Tosh ! " (Note by R. B. Peake.)

The party presently stayed at Mulgrave Hall. A spirit of active research predominated over all the amusements, inspired by the two leading members of the party. Botany and opening ancient barrows were the chief objects. Banks put the two boys into active training as botanists, and sent them out early in the morning to the woods to gather plants. " We could not easily have met with an abler master. Although it was somewhat early for us to turn natural philosophers, the novelty of the thing, and rambling through wild sylvan tracts of peculiarly romantic beauty, counteracted all notions of studious drudgery, and turned science into a sport. We were prepared over-night for those morning excursions by Sir Joseph. He explained to us the rudiments of the Linnean System, in a series of nightly Lectures, which were very short, clear, and familiar ; the first of which he illustrated by cutting up a cauliflower."[1]

After leaving Mulgrave Hall the party went on to Skelton Castle. In the adjacent village of Kirkleatham there lived the father of Captain Cook, who excited much interest in his visitors. " His looks were venerable from his great age ; and his deportment was above that which is usually found among the lowly inhabitants of a hamlet. . . . His eightieth year had nearly passed away, and only two or three years previously he had learned to read, that he might gratify a parent's pride and love by perusing his son's first voyage round the world."[2]

The return of Captain Cook from his second voyage round the world was honoured in no niggardly way. To the general public he was a hero, a seaman who was carrying forward the best traditions of his profession. The Admiralty officials were not less sensible of his value to the service. The Royal Society admitted him to the Fellowship, and listened eagerly to his paper read before

[1] R. B. Peake : *Memoirs of the Colman Family*, I, 355, etc.
[2] *Random Records*, by George Colman the younger, I, 202.

the Society, in which he detailed the measures taken to secure the health of his crew and the success with which he had fought the demon Scurvy. For this paper, moreover, the Royal Society awarded the Copley medal, on an occasion marked with considerable enthusiasm by the members present.[1]

It soon became clear that the British Government were anticipating Cook's second return, only to send him forth on a still more important voyage, viz. the possible discovery of the North-west passage to India.

Banks does not appear to have had much direct concern with the new project, beyond friendly consultative talks at the bidding of Lord Sandwich. Yet his personal interest in the affair was extensive, for most of the officers, including Cook himself, were to some extent intimates of Banks. Cook accepted the commission without hesitation. There was William Bligh on board as Master. Lieutenant James Burney, Lieutenant Clerke, Molesworth Phillips, lieutenant of marines, Dr. William Anderson the surgeon, who acted as naturalist,[2] and others who had proved very creditably their promise of capacity for the arduous service in view. Also a Kew gardener, David Nelson by name, was taken on board at the instance of Banks.

The ships, *Resolution* and *Discovery*, got away in July, 1776. Letters dated November, from the Cape of Good Hope, reached Banks from Dr. Anderson and from Captain Cook.[3] They were never again in touch with

---

[1] " If Rome (said the President, addressing his fellow-members) decreed the Civic Crown to him who saved the life of a single citizen, what wreaths are due to that man who, having himself saved many, perpetuates in your ' Transactions ' the means by which Britain may now, on the most distant voyages, preserve numbers of her intrepid sons, her mariners: who, braving every danger, have so liberally contributed to the fame, to the opulence, and to the maritime empire of their country."

[2] According to Cuvier, Captain Cook had refused to take with him an independent professional Naturalist.—*v. Éloge* on Sir J. Banks.

[3] After all, the old *Resolution* seems to have been a matter of grave concern to the captain. He writes : " If I return in the *Resolution*, the next trip I may venture in a ship built of gingerbread."

E

civilization until May, 1779, when they enjoyed the
hospitalities of the Russian Governor at Kamtchatka.
Soon the news reached England of the death of Captain
Cook: "murdered by the Indians of an island where
he had been treated if possible with more hospitality
than at Otaheite," according to Clerke's dispatches
homeward, dated January, 1780.

Lieutenant Clerke, also, did not live to see again
his native land. He wrote to Banks (August, 1779)
saying that his health was broken, and that his "stay
in this world must be of very short duration"; and
that he recommended certain of his shipmates to Banks's
notice as deserving of his protection in case of need.
Clerke subscribes himself "your devoted, affectionate,
departing servant." He died a few days after this;
one of the very numerous men who had been strongly
attracted by Banks's personality.

Clerke's dispatches reached Lord Sandwich early in
1780. He was inexpressibly shocked at learning the
fate of his commander, and hastened to Banks to distress
him in turn. Alike among their friends, the naval service
and the scientific world, the news of Cook's death caused
an outburst of genuine feeling. Even beyond the confines of his island home there were thousands to regret
his loss. The French people particularly recognized his
merits; and they have never ceased to honour his memory.
The British Government placed his widow beyond the
risk of want by granting her a pension; which, by the
way, she enjoyed for the long period of fifty-five years.
The Royal Society determined that special honour should
be paid, on occasion of this tragic loss of one of their more
distinguished members; a medal was struck bearing a
profile of Captain Cook and a reverse of Britannia pointing
toward the South Pole. The subscription toward the
expense of this was so generous that the thing was done
in most noble fashion. Fourteen gold, two hundred and

eighty-nine silver, and five hundred bronze medals were struck for distribution.

The activity of editors and booksellers was naturally stimulated by the return of the circumnavigators. This third Expedition promised to be of certainly not inferior interest. As it happened, there were startling novelties about it; and the tragic fate of the Commander was so unexpected and so grievous that the whole nation took it to heart. People had formed a high notion of Cook's character. He was persevering and full of resource ; a first-class seaman and nautical surveyor ; a strict disciplinarian ; he kept watch over the health and morals of his subordinates; and his uniform success in dealing with savage people, and making friends with them, cannot be said to have been broken in the final catastrophe. Hence the honour paid to his memory, and the general wish to know all the proceedings of the last years of his life.

The official account of this last voyage was put into capable hands. Lord Sandwich and Mr. Banks were minded to have the thing well done, and they personally superintended during its preparation. Even George III was called into consultation. Dr. John Douglas (afterwards Bishop of Salisbury) took the literary part of the work, while Captains Gore and King and Lieutenant Burney contributed their journals and their personal recollections.[1]

The result was a fine book in three volumes, which, however, did not appear until 1784. The profits of its

[1] In the MS. Dept. of the British Museum (Addl. 8955) is a transcript of Burney's Journal, written in a microscopic hand and enclosed in a tiny case ; the weight of the paper only half an ounce. A memorandum on the enclosure states that, " On the arrival of the *Resolution* and *Discovery* in Macao Roads, we learnt that Great Britain was at war with France and with the North American States, which gave us some apprehension of being captured on our passage homeward ; and in such event that we should lose our journals. Under this apprehension, I made a copy of my journal on China paper, in so small a compass as to be easily concealable ; that if bereft of our journals there might be one saved for the Admiralty."

sale were very great, and Mrs. Cook's share provided her with a fair annuity.

Meanwhile, one year after another discovered a new candidate for the honours attaching to the story. Ellis, the surgeon, produced the earliest. He was an artist also, and his drawings made excellent illustrations. As it is written well, and it probably had a good vogue as the first published account of the voyage, it ought to have been a more profitable venture than it was. Only there was the poor bookseller to be considered.

It appears that Banks had duly befriended Ellis (on Lieutenant Clerke's suggestion), and had advanced him money. For some reason or other, perhaps because he had imprudently run to a publisher without consulting Banks, Ellis found himself compelled to offer apologies and regrets. He was very poor, and, " full of expectation, called upon the bookseller." After much haggling, he consented to take fifty guineas—the only alternative being " half profits." His very long letter to Banks is an explanatory apology, upon " hearing " that he has incurred his displeasure. It is very sad to read, and too long to print here. But Banks's reply is really deserving the reader's attention, especially the last paragraph.

Sir Joseph Banks to W. W. Ellis.

" SOHO SQUARE, *January* 23, 1782.

" SIR,—I received yours, and am sorry you have engaged in so imprudent a business as the publication of your Observations.

" From my situation, and connexion with the Board of Admiralty, from whence only I could have hoped to have served you, I fear it will not in future be in my power to do what it might have been, had you asked and followed my advice.

" Your note of hand, which is in my possession, I am

ready to deliver up to you, and to pay you £30 more on account of your drawings, which I value at £50, whenever you choose to consent to give me a receipt for that sum as their price ; as I do not think it proper to insist upon the right to them which I derive from your having made me a present of them, or while your circumstances are so much confined.

"The only service you could do me would be by entering into a plan to revenge yourself of the bookseller for his Judaic treatment. If you would heartily join in it I would assist. Your humble servant," etc.

About this time there appeared at Berne (in French) and at Göttingen (in German) *Voyage autour du monde avec le Capitaine Cook, par Henri Zimmermann.* There is no copy in the British Museum. But the title appears in Boucher's List. It is a curiosity, because the name does not occur anywhere except in a letter from the Rev. Michael Lort to Rev. George Ashby (May 4, 1784), yet ought to be of great interest :

" When abroad last summer I bought a French account of the voyage by one Zimmermann, a Swiss who was on board the whole time. Sir Joseph Banks, to whom I lent it, was much pleased with it, and told me it contained some curious particulars not in the larger work."

John Ledyard followed, with a *Journal of Captain Cook's last voyage to the Pacific Ocean,* published at Hartford, Connecticut, 1783.

David Samwell, surgeon of the *Discovery,* essayed a description of the last voyage ; but only part of it was published : *A Narrative of the Death of Captain Cook. . . .* London, 1786. The remainder of the work is still in MS. (Egerton MSS., 2591.)

Lastly, there is an unpublished MS. in the British Museum of *A Journal of a voyage undertaken to the South Seas on Discoveries, by His Majesty's Ships " Resolution " and " Discovery," Captain James Cook and Charles Clerke, Esq., commanders.   Kept by Thomas Edgar, Master.*

This story is preceded by an account of the ship *Resolution ;* how it was built at Whitby, and was a year and a half old when bought into the service in 1776. (Additional MSS., 37528.)

George Forster was commissioned to do the German translation. He was now at Cassel, having found a respectable Professorship, which afforded much better prospects than he seemed able to get in England. Although he had returned to the Fatherland, Forster was greatly attached to this country   He wished to honour its people to the best of his power.   From his correspondence it is clear that Banks appreciated all this, and continued a fast friendship with a man of real merit. He sent to Forster early sheets of Dr. Douglas's book. In the event, a good German story was produced, which was as popular, perhaps, in the Fatherland as the topic was on this side of the water.   Though Germany at that time had few or no sailors, they frankly appreciated the great credit that Englishmen were deriving from the exploits of their seamen.   Forster mentions, under date January 10, 1781, that he had a very ample detail of the late voyage, from the mouths of two Germans who made the voyage as common sailors.

CHAPTER IV

PRESIDENT OF THE ROYAL SOCIETY

THE President of the Royal Society, at the beginning of the year 1778, was Sir John Pringle, M.D.

Dr. Pringle was from Roxburghshire, born in 1707, educated at St. Andrews, Edinburgh, and Leyden. He settled in practice at Edinburgh, became Professor of Moral Philosophy, and soon attained a high reputation. In 1742 a new career opened for him, in serving as Physician with the British Army. For three years he was in attendance on the hospitals in Flanders. He was present at the battle of Dettingen, and saw more or less of constant military service until 1749, when affairs permitted him to establish himself in London. He had been elected a Fellow of the Royal Society in 1745 ; read several papers which were printed in the *Transactions ;* and in 1753 became a member of the Council. He was now at the top of his profession, and due honours fell upon him. He was made a Baronet, and appointed physician to King George III ; and in November, 1772, attained the distinction of being elected to the Presidency of the Royal Society.

After six years, during which he continuously upheld the dignity and the traditions of the Society, Dr. Pringle's health began to fail. He was now over seventy years of age, and felt it was time to let the controlling affairs pass into younger hands. At the anniversary meeting of 1778, he announced that his time was come for retiring.

Blessed with a very wide circle of friends, he was enabled to enjoy his few remaining years in comfort and honour, and still contributing to the promotion of Science. He died in 1782.

Several names were mentioned of possible successors to Sir John Pringle. Others beside that of Banks had the honour of being included among them. As time went on, it was apparent that the only serious candidature beside that of Banks was on the part of the friends of Mr. Alexander Aubert, a wealthy merchant, and a very active member of the Society, elected in 1772. It was afterwards asked " what Mr. Aubert had done," a question, however, that has often been put under similar circumstances, concerning men superlatively active in good work yet disinclined to seek public renown. Aubert was best known in the scientific world as an astronomer.   He built three observatories : one at his house in Austin Friars, another at his residence at Highbury, and a third near Lewisham. And, from casual references to him, it is quite clear that he was a greatly respected man in scientific circles, genial and hospitable in private life, and an important member of the Royal Society.

Dr. Solander to Sir Joseph Banks.

" LONDON, *August* 11, 1778.

" MY DEAR SIR,—This morning Mr. Planta told me that Sir John Pringle has certainly declared that he intends to resign ; and Mr. Cavendish says that Sir John has mentioned it at the *Mitre*. It is true that he has given hints about Mr. Aubert, but all look to you. Dr. Pitcairne and others have desired me to tell you that."

Mr. Weld says (*History of the Royal Society*) that a general opinion prevailed that no one was better qualified to occupy the vacant chair than Mr. Banks. It would seem, from this note of Dr. Solander, that the opinion was prompt and spontaneous, as it is given so early after Dr. Pringle's announcement. By the time, however, that the opening of the winter session approached, several other names had been suggested for the vacancy. There was arisen an anti-Banks party among Dr. Pringle's friends ; and it appeared there would actually be a contest for the vacant Chair. Banks wrote to Richard Gough, November 23, asking his assistance in the candidature,[1] and doubtless thought proper, once a point of rivalry being admitted, to urge his personal friends to give him their support. There can be no doubt, the question having been raised so largely in his favour, that Banks would be inclined to believe it a laudable ambition to succeed Dr. Pringle. For a masterful spirit like his, that inclination would soon resolve itself into an opinion that the Society could not do better than choose him, since so many of the members were in favour of the idea.

Sir John Cullum to Rev. Michael Tyson.

" *December 7, 1778.*

" Mr. Banks was elected President, unanimously to appearance by 220 votes.

" There were 127 of us at dinner, among whom were several of the Nobility. The dinner was late, owing to the new President, who waited for the declaration of the new Secretary, which was Mr. Maty, by a majority of two to one. . . . The President came in a great hurry, quite out of breath, and sitting down (I was opposite to him) said with good humour, but with rather too little dignity : I believe never did a President of the Royal Society run so fast before.' However, his behaviour throughout was very proper. . . . "

---

[1] Nichols : *Lit. Illus.*, IV, 693.

It soon came to be seen that the right man had been chosen. A good-natured autocrat was wanted. The unique position occupied by the Royal Society required a system of government calculated to preserve its authority in matters of Science. None could fulfil the duties of President but one already popular from his known character; and the authority of the Society happily blended with the authoritative masterfulness of Banks. It was a triumph of the " one man " system. All over Europe the Society now became scarcely distinguishable from its President. There was an internecine disturbance, in the year 1784, to which reference will presently be made, which only made Banks's position firmer; and from that time to the day of his death his personality was overwhelming. Not that he was overbearing (although one might think differently who consulted the caricaturists, and the unhappy sufferers from envy); on the contrary, there was no rule asserted, neither was any refusal given, except in terms of perfect courtesy. There will be plenty of evidence to this effect in some of the occasional letters included in the following pages.

The Royal Society was now become a venerable National institution. During a career of eighty years or more a system of recording the experiments, and the philosophical speculations, of its members had been followed, which resulted in placing every branch of National Science on an intelligible footing. At the period of the foundation of the Society, any organized scientific pursuit was almost unknown. The lover of philosophical experiment performed his labours in the dark; unless he cared to run the hazard of persecution, public or private. The power of ecclesiastical censure lingered in a subdued form long after the Churches had become professedly free. Nature and her latent powers were *tabu*, as matters of inquiry, and every manifestation of unwonted sort remained the object of superstition and credulity.

It was during the fifth and sixth decades of the seventeenth century, when extremest licence of thought was permissible if only it were shielded from exposure and controversy, that intelligent men sought freedom from the vexatious troubles of the time by engaging in scientific inquiry. As Dr. Whewell finely says : "There arose about this time a group of philosophers, who began to knock at the door where Truth was to be found." [1] After the Restoration, the numbers of these pioneers increased. Their meetings were frequent, and well-attended. On July 15, 1662, a Royal Charter was granted them, constituting them a body corporate as the Royal Society. The King himself was a devoted patron ; his office was not purely nominal, for he took an active and often sagacious part in the proceedings. He was elected a Fellow on January 9, 1665, being one hundred and seventy-first on the roll.

" The business and design " of the Society is the subject of a quaint document still in preservation at the British Museum.[2] It is too long to quote here in full. But the gist of it is fairly well contained in these two sentences :

" To improve the knowledge of naturall things, and all useful Arts, Manufactures, Mechanick practises, Engynes and Inventions by Experiments (not meddling with Divinity, Metaphysics, Moralls, Politicks, Grammar, Rhetorick, or Logick)." " In order to the compiling of a complete system of solid Philosophy for explicating all Phenomena produced by Nature or Art, and recording a rationall account of the causes of things."

On such safe basis the Society began its useful labours ; and to these principles it has unswervingly adhered to the present day.

Animal and vegetable physiology were the most attractive subjects to people of an inquiring turn of mind. Hence the earlier *Transactions* are largely occupied

[1] *History of the Inductive Sciences*, II, 145.   [2] Addl. MSS., 4441.

by these and allied topics; as Anatomy and Medicine, Zoology, Agriculture, and Systematic Botany. Mineralogy, and the unaccountable *lusus naturæ* which later led to systematic Geology, came into prominence from the first. The Mathematical Sciences, with optics and astronomy, occupied the minds of a large section of inquirers. Electricity and magnetism had a few students and observers soon after the opening of the eighteeenth century. Chemistry came later to the front. Not all these things were gone through without much stumbling. But there was little that was done that did not manifest zeal and earnestness, and a sound belief in the destiny of the Society.

So, in eighty years of assiduity, and real love for their labours, the members of the Royal Society had founded systems in every branch of cotemporary Science. They had set a basis for further study which was of incalculable value. The Society had attained a world-wide fame, which attracted the most illustrious European physicists and naturalists to its fold. Its stability was perfect; a circumstance unquestionably due to adherence to its own first principles.

Mr. Pennant was one of the earliest among distant friends who sent congratulations to Banks on his accession to the Chair of the Royal Society. He writes warmly from Downing (December 14) to this effect; and proceeds with, " Let me add my wishes that something like Natural History may appear under your auspices in the annual productions of the Society."

The following extract concerns us as evidence of the notoriety of Banks's name all over Europe. The writer is Jean H. de Magellan, a gentleman whose residence alternated in Paris and London. He was a genuine product of the eighteenth century. He had been a monk, but a studious one; and an ever-opening mind forced him to reject the monastic career. He delighted in associating with scientific men ; and being a good linguist, and having travelled over Europe as a tutor, was well equipped for such company. His work lay chiefly in perfecting optical and mathematical instruments. Having joined the Royal Society in 1774, he proved a very useful member. Magellan writes (December 28) : . . . I have received many congratulations from my friends abroad, on account of your election of President, and particularly from Messrs. Mann, Needham, Baron de Poederle, etc. This last came to London together with the late Duke of Arenberg, and I accompanied them to your house in Burlington Street, a little after your circumnavigation with Captain Cook. You made a present of various curiosities to the Duke, which he deposited in the cabinet of Prince Charles at Bruxelles. The following are the words of the first, viz. Mr. Mann : ' Notwithstanding my sincere regard and respect for Sir John Pringle, whose civilities to me in London I shall never forget, I must own that Mr. Banks's election to that Illustrious Chair gave me the most sensible pleasure and satisfaction. I esteem him, and love him from my heart, and in return he honours me with a particular friendship, which I will do my best to cultivate. His character and great celebrity will do honour to the Society.' "

Sir William Hamilton to Joseph Banks.

" CASERTA, *February* 9, 1779.

" DEAR BANKS,—I have been these two months in pursuit of Boars, Wolves, Foxes, Ducks, Woodcocks, and Snipe, with His Sicilian Majesty. The sum total of the chase in last month is really frightful : 6922 pieces. The King is the coolest and best shot I ever knew.

" Your election to the Presidency of the Royal Society gives universal satisfaction, and me a very particular one, Mr. President.

" . . . I have sent Solander a collection of corals for our Museum. There are duplicates of many, which he may give to you if they should be curious and worth your acceptance."

Writing again (March 23), Sir William relates some more sporting scenes, with enormous bags ; and adds the following : " Yesterday the King of Naples sent a spaniel puppy of about four months old to shew me as a curiosity. Out of the pupil or centre, of his left eye grows a tuft of soft hair like that of his body ; about an inch long, and it seems to be destroying the sight, which is much impaired. As I never saw or heard of anything of the kind I thought, Mr. President, it was my duty to communicate this phænomenon to you."

Banks was married in March, 1779. The lady was Dorothea, eldest daughter of William Western Hugessen, Esq., of Provender, in Norton, Kent.[1]

Banks had taken the large house in Soho Square, which remained his principal residence for the rest of his life, in the autumn of 1777.[2] It was become the resort of all persons, and all classes of students, who shared with Banks those philanthropic principles which aimed at ameliorating the lot of Humanity by endeavours to disperse the mists of Ignorance. Every one was welcome who, by improvements in the Arts or by further unravelling the secrets of Nature, had any scheme for the benefit of his fellow-men. Young authors were welcome who wished to consult the splendid library.

---

[1] *Rev. Sir John Cullum to Rev. George Ashby* :—

" May 22, 1779. . . . Yesterday morning I took a breakfast with Mr. Banks, who told me he was always glad to see his friends at that meal. And when can one see him so well ? For after breakfast he retires into his study with those that please to attend him ; where those who are likely to visit him will meet with ample entertainment. His wife is a comely and modest Young Lady.

" Solander was particularly cheerful and talkative."

[2] No. 32, Soho Square, now occupied as a Hospital.

LADY BANKS   SIR JOSEPH BANKS

From a Wedgwooa cameo, attributed to Flaxman

Foreigners and strangers were welcome ; and although suitable introductions were rigidly exacted, these helped to swell the crowd of inquiring visitors. The house was a vast museum ; in which books, pictures, rarities from all parts of the world, and innumerable botanical specimens, delighted the varied company, and gave magical effect to Banks's frequent *soirées* and less formal gatherings. Over all these things Dr. Solander reigned as Librarian and Curator ; while Banks's sister, who was devoted to him, had been mistress of the house. Miss Banks was quite a lady of fashion at this time.

Two years later, in March, 1781, Banks was honoured by the King with a baronetcy.

Among the crowd of persons that thronged the rooms at Soho Square at this period was William Herschel. He was known to a few friends as a man devoted to Science. He had already contributed papers to the *Philosophical Transactions*. The discovery of the planet which was at first called *Georgium Sidus*, and his application to the improvement of telescopes, suddenly and dramatically put him into " the full blaze of fame." Herschel was of a singularly modest disposition ; and his researches would have been hampered through pecuniary difficulty but for the generosity of his fellow-philosophers. Hornsby and Maskelyne, and the other astronomers, far from displaying jealousy, recognized Herschel's merit and lent themselves to the support of his claims for distinction. Dr. William Watson, his neighbour at Bath, wrote to Sir Joseph Banks enthusiastically in praise of the man and of his new telescope ; and further urged that, as the King had heard of him and was deeply interested in the new discoveries, some appointment at Court would be a suitable way of helping Herschel ; thus providing him with an income which would make up for the loss of that which he had earned in his musical profession.

Thus prompted, Banks forthwith determined to use

his personal influence with the King. It was here that Sir Joseph was often so helpful when there was a meritorious object in view. George III had acquired much faith in Banks's judgment generally, and was now an intimate friend, prepared to accept his counsel in all matters of scientific advancement. His Majesty readily found excuse for granting Herschel an appointment worth two hundred pounds a year. The astronomer was presently received at Court, and had the delight of showing his newest instruments to the assembled Royal Family. A long and close friendship continued between Banks and Herschel.[1]

The circle in Soho Square was rudely broken one morning in the spring of 1782. Dr. Solander was seized with paralysis just after breakfast, while in conversation with Blagden. The best physicians were speedily at hand, while Lady Banks made immediate provision for him to remain in the house. " You may judge of the affliction of every one here," writes Dr. Blagden to Sir Joseph. All the efforts made for the patient were in vain, and he died in the course of a few days.

The death of Solander was a serious loss to the Banks coterie. He made friends with all. His habitual presence among the company to be seen in Soho Square gave lustre to the scene, alike in his attractive disposition and in the graces of learning and intelligence. Not only was he Banks's right-hand man in his scientific work, but, from the day when Banks invited Solander to go round the world with him, they had been bound in firmest friendship. And he was always acceptable in good

---

[1] Fanny Burney found Herschel a constant charm. He was often in her tea-room at Windsor, with Dr. Lind and others of her little circle. He was " openly happy in the success of his studies " . . . " a delightful man ; so unassuming with his great knowledge, so willing to dispense it to the ignorant, and so cheerful and easy in his genial manner." He could play sweetly on the violin, and vary his entertainment by "shewing me some of his new-discovered universes." Some pleasant glimpses of Sir William Herschel are to be found in Madam D'Arblay's *Diary*.

society. Samuel Johnson found much pleasant resource in Solander's company; and the great man's more fashionable friends, too, made him very welcome.

Daniel Carl Solander was a friend and pupil of Linnæus. He was an apostle of the "*Systema*," and had much to do with making that philosopher's works familiar to the people of England. Soon after his arrival in this country, in 1760, he was employed at the British Museum; and, in one way or other, was connected with it until his death. His early friendship with Banks doubtless began there. In 1764 he became a Fellow of the Royal Society. After his return from the voyage in the *Endeavour*, the University of Oxford conferred upon him the honorary degree of D.C.L. He left behind him a large collection of Botanical MSS., which are now preserved in the Natural History Museum. Beside Botany, Solander had studied conchology assiduously. Since 1779, he had been acting as Curator at Bulstrode House, in charge of the splendid collection of the Dowager Duchess of Portland.

Dr. Jonas Dryander, another learned Swede, already established in the intellectual society of London, and well known to Sir Joseph Banks, henceforward filled the post vacated by the death of Solander. He was a rather more active man, and became as valuable an adherent of the Banks establishment as his predecessor.

John Christian Fabricius [1745–1807] was another clever Scandinavian naturalist who assisted Banks at this period. He had been a pupil of Linnæus, and associate of Thunberg and others. He became noted for a new classification of insects, in advance of former ideas. In his turn he was superseded; but he performed great services to natural science.[1] Fabricius was in England

[1] Among other writings he is remembered for *Systema Entomologiæ* (1775), and *Briefe über England*, etc. (1783). For a notice of his work, *v.* J. G. Children, in *Philosophical Magazine* for February, 1830, p. 118.

F

1767-8. He made acquaintance with Solander, and through him was introduced to the best scientific society in London. Banks, Hunter, Fothergill, John Ellis, Pennant, Greville, were among his intimate friends. In the years 1772-5 he spent his winters in Copenhagen and his summers in London. He lived pleasantly in Banks's company, spending much time over the treasures which had come from the South Seas. After 1780 he was again in London, delighted with the further rich accumulations in Banks's house. The insects were put into his hands entirely. The store of specimens which ultimately found a home in the British Museum owed its perfection to the labours of Fabricius.

The Duchess of Portland's wonderful collection of objects in nature and art was deservedly renowned by her cotemporaries. When the treasures were disposed of by auction, in 1786, the sale occupied thirty-eight days! Readers of Mrs. Delany's *Autobiography* will remember the almost tiresome zeal of everybody concerned with this museum. Lightfoot, the botanist; Solander; Jacob Bryant, the antiquary; Ehret, the most admired painter of flowers and foliage, in turn were serving the Duchess in her pursuits. Mrs. Delany herself enjoys it all. " We have in attendance Mr. Ehret . . . who goes out in search of curiosities in the fungus way, as this is now their season [Sept., 1769], and reads us a lecture on them an hour before tea, whilst her Grace examines all the celebrated authors to find out their classes. This is productive of much learning and of excellent observations from Mr. Ehret." [1]

[1] Georg Dionysus Ehret, b. Heidelberg 1708, was in the garden of the Elector of Heidelberg. He travelled great distances over Europe, botanizing and drawing. After coming to London, he made friends of Hans Sloane and other gentlemen. Became noted for his perfect delineations of plant life ; and gave lessons in drawing to the highest personages. He was much appreciated by Banks, both as friend and draughtsman. A great many of his beautiful drawings are housed in the Botanical Department of the Natural History Museum : well worth a pilgrimage to any lover of this branch of art.

Mrs. Delany, too, succumbed to the mania. In the Print Room of the British Museum is a large assembly of imitation plants, the work of this lady, which will astonish any curious visitor who cares to apply for a glimpse of them.[1] These things were the admiration of Banks and Solander. Indeed, Sir Joseph declared that Mrs. Delany's representations of flowers were the only imitations of Nature, that he had ever seen, from which he could venture to describe botanically any plant without fear of committing an error. A description of Mrs. Delany's method will be found in her *Life and Correspondence* (VI, 96, 97).

Jacob Bryant was another very welcome visitor at Bulstrode House ; an antiquary, with very wide views and an extensive knowledge of history. Antiquities were, in his eyes, the handmaid of history ; perhaps the first man who thus devoted himself to the topic. He wrote many books, and collected a great number of gems, coins, etc. Apparently an insignificant man, who would be overlooked in the smallest crowd, he bore the highest character for goodness to his fellow-men and for his extensive learning. Through Mrs. Delany and Fanny Burney, several of the Duchess's friends were introduced to the homely Court at Windsor. Bryant was frequently in Miss Burney's tea-room, and the King became very friendly with him, visiting him at his house at Farnham Royal and staying long in his company.[2]

Another friend of Banks to be found in the Windsor coterie was James Lind, physician, traveller, astronomer. In Miss Burney's time, Lind was resident in Windsor ; in good practice, beside being Physician to the Royal Household. He was one of Banks's oldest friends, and

---

[1] A printed list is to be found in the Banksian Library : *A Catalogue of plants, copied from nature in Paper Mosaic, finished in the year* 1778, *and disposed in Alphabetical Order, according to the generic and specific names of Linnæus.*—Press mark B $\frac{100}{2}$.

[2] *v.* Madame D'Arblay's *Diary*, vols. III, IV *passim*, for some anecdotes of Bryant.

a very charming person. He was respected everywhere in scientific circles, at once for his attainments and the modesty with which he carried them. He was one of the party who went to Iceland with von Troil. He had been to India and back, as surgeon on board an East Indiaman. It had been proposed to send Dr. Lind on Cook's third voyage. He made it conditional on Banks also going in company, " from the real regard I have for so able and excellent a man." They had been somewhat ungraciously treated in 1772, when both Lind and Banks went to great trouble and expense in preparing for the second expedition. The latter proposed to reimburse Dr. Lind, who says (in a letter to Dr. Maskelyne) : " He told me that he had a good estate ; his estate he looked on as belonging to his friends as well as himself ; that he held me as one of them, and begged me to command my share of it whenever I wanted it."

CHAPTER V

THE ROYAL SOCIETY—*continued*

IT was in the summer of 1783 that Sir Joseph had a first serious warning of the common lot in life. Hitherto he was active, and buoyant, as a man at forty years of age should be. There is no hint of his having yet had a day's illness. But there was a carriage accident in August or September, the result being that he was laid up for some weeks at Revesby, an invalid. After that his habits were never the same again. The demon Gout insidiously began to lay hold upon him, and eventually troubled him as long as he lived.

During Banks's confinement indoors, his friend Blagden was more assiduous than ever in his letters from town, and entertained Sir Joseph with gossip of all sorts. When this worthy gentleman had the opportunity, he could pour forth long messages with extraordinary verbosity. His letters during this particular autumn are a rich chronicle, worthy of the notice of any student who would reproduce an idea of the doings of the time.

So, Dr. Blagden made a good newsman for Revesby, with his stories of flying-machines, Franklin's electrical experiments, Herschel and his telescope, his own speculations on the composition of air, beside dozens of other matters which were keeping the philosophers wide awake. All these things, together with minor gossip of a personal nature, doubtless softened to the invalid the loss of his wonted share in affairs.

Nor did Dr. Blagden omit due service to Lady and Miss

Banks. " . . . The songs in ' Gretna Green ' are now published, and I shall send a copy of them by the Horncastle coach next Monday Morning. Most of them, I see, are old. Whether the newer ones have any merit the ladies must decide. I add the last collection of Vauxhall songs, as the first air in it is said to be a great favourite." ". . . . The ladies do me too much honour in condescending to take notice of any poor attempts of mine for their amusement. The music went on Monday." ". . . I have just seen your coachman ; who says he is now well, and seems so. Your own hurt, though I hope from the free intervals it leaves you all bad consequences will be averted, yet is sufficiently alarming to render caution very necessary. The intensity of exercise is undoubtedly of more consequence than the quantity, though it is very possible for you to exceed even in the latter respect."

Then there was a sad tragedy of a turtle, which seems to have weighed upon the Doctor's mind. He feared it could not be saved for a feast, at the head of which Sir Joseph would preside.

" At the *Crown and Sceptre*, Thursday, I asked advice of Simpkin the landlord, relative to your turtle, and begged him to call and see it. He gave but a bad report, being of opinion that it was not likely to live many days, and the weather was much too cold for keeping it in water, at least above half an hour in the day ; and that it should be suffered to travel about the kitchen, and be wrapped up in a blanket at night." In this condition Mr. Simpkin offered five guineas for it. But it was determined by the Club to have it killed. " Your turtle was dressed this day at the Crown and Anchor. Simpkin, having cut away all the suspicious parts, made three tureens of soup, and no other dish. It was very good, and well cooked ; but not to be compared with a plain turtle steak or cutlet. There was but a small company ; all who were present, however, expressed much obligation to you for sending it."

THE ROYAL SOCIETY 71

About this time all Europe was diverted with the new and wonderful experiments in aerostation. Sir Joseph Banks, in common with most of the leading philosophers of the day, took immense interest in them. It is not certain that Banks financed many of these essays of the balloonists. But, obviously, there were attempts made to get him to do so. It is possible that Lunardi, if any one, was successful in tapping the financial resources of Soho Square. Jeffries, from America, who made the first voyage across the English Channel, probably gained Sir Joseph's patronage ; since he wrote from Guines an exultant story of his exploit and his reception by the municipality of Calais. Blanchard carried on an Aerostatic Academy, and was frequently engaged in conveying parties of adventurous English gentry, with more or less success, for trips toward the sky. There was Henry Smeathman, who had done some work in Africa in Natural History, and he kept in touch with Banks ; invented a steerable balloon, and persisted in hoping for Banks's conversion to his plan after having been assured that there was too much hazard about it.

Dr. Blagden's letters of the period are full of ballooning adventures, and observations in company with Aubert, Dryander, and others, who mounted roofs and observatories with sedulous zeal in the philosophic side of the question

The public craze lasted in some vigour for two or three years. There seemed to be no bound to its possibilities, after development on the lines of inflation by a gas lighter than common air. Naturally, the scoffer was in force. Ballooning lent itself easily to caricature. And there were many found to defy the tremendous innovation, on the ground that its utility was yet to be discovered, and it was only fit to " gratify idle Curiosity and promote Dissipation." Yet the sport was undeniably popular, as it has never ceased to be. And the character of the philosophic men who continued to promote it was too high for the thing to be crushed by premature disapproval.[1]

The session of 1783–4 was disturbed by a crisis in the affairs of the Royal Society, the most momentous in its history.

The popularity of Sir Joseph Banks as President of the Society had two sides to it. Occupying that distinguished post, he was recognized as a worthy and most zealous patron of Science, devoting his wealth and his energies without stint to the welfare of the Society. All this was very excellent and commendable. But, beside this, he was a masterful man, of great sagacity, dealing with matters in a rough-and-ready way ; indisposed to permit the needless interference of side issues with great principles. Now, if this quality of strength is one that the average Briton treasures above all, not the less is he apprehensive of any tendency to dictatorship. And it happened that, during the early years of his occupancy of the Chair, there were to be found Fellows of the Society who held that Sir Joseph was occasionally too self-assertive.

One thing followed inevitably upon the accession of Banks to the President's Chair. He determined to restore the older restrictions as to candidature for membership. He would have the Royal Society so far exclusive as to forbid the admission of such gentlemen who seemed to regard the distinction mainly as a step in personal advancement. The men who sought to enter the Society should be those who had distinguished themselves in some one or other branch of Philosophy, and those of

---

[1] v. Gentleman's Magazine, and the newspapers, and Miss Banks's Collection of Broadsides in the British Museum, for coloured pictures and views, for 1783–5; and much anecdotage on the topic. There is room here for one story. At Constantinople, the people who saw a balloon come down cried out with fear. " They were seized with inexpressible horror, thinking it was the Prophet come down to punish them for their sins ; and they fell prostrate on the ground."

good social position who were willing to expend their wealth and influence in the promotion of Science. Some latitude had grown up in this way ; a vital point if the Society would preserve its dignity and renown.[1]

In the course of devoting himself to the interests of the Society, Sir Joseph soon made his personality felt ; and, among other things, "took measures to render the Fellowship more difficult of attainment than it was at the time of his election." The Secretaries, there is reason to believe, had the power of electing any candidate who was ambitious of becoming a Fellow of the Society ; and the President, whose " duty it clearly is to preserve the purity of election, was seldom consulted."[2] Banks presently announced to the Secretaries and Members that he meant to watch over the application for admission. " Previous to an election, he spoke to the Members who usually attended ; he gave his opinion freely on the merits of candidates, and when he considered a rejection proper, he hesitated not to advise it, giving his opinion and recommending or asking a black-ball from individuals at the time of the ballot."[3]

There were eleven rejections in the first five years of his incumbency. Hence the partial loss of popularity which the President suffered. In the year 1783 discontent came to a head, doubtless fomented by the Secretaries.

Mr. Paul Henry Maty (b. 1744 ; son of Dr. Matthew Maty, a distinguished physician of Huguenot parentage) was the General Secretary, which office he had held since 1778, having previously been Foreign Secretary for six years. With many respectable literary friends, and a

[1] A humorous story was afloat at one time, bearing on the facility with which admission was granted to the Royal Society : " D'Alembert used jocosely to ask any of his acquaintance coming to England, if they wished to become members of the Society ; intimating that, if they thought it an honour, he could easily obtain it for them."—*Tilloch's Philosophical Magazine*, vol. LVI.
[2] C. R. Weld : *Hist. of the Royal Soc.*, II, 152.
[3] Brougham, *op. cit.*

somewhat energetic disposition, he found a congenial resource in the Royal Society, to which he was elected in 1771.

Mr. Joseph Planta (born in Switzerland, 1744) was the son of a German Reforming Pastor in London. He was assistant, and afterward chief, Librarian at the British Museum for the long period of fifty-four years, with entire public approval and benefit ; a leading figure in the intellectual life of London. He had been Foreign Secretary to the Royal Society for two years, and in 1776 became Second or Assistant Secretary.

It appears from Blagden's correspondence with Sir Joseph that both Maty and Planta were aggrieved by the firm demeanour of the young President. They had raised a strong party among the members of the Society, opposed to the continuance of what they held to be a too highly pitched notion of presidential authority. The storm burst suddenly upon the Society at the opening of the session of 1783, and arose out of the allegation that Dr. Hutton, the existing Foreign Secretary, was not performing his duties efficiently. Now, Charles Hutton was truly a busy man, one of the first mathematicians of the age. He had been chosen Professor at the Royal Military Academy at Woolwich by a very strong Committee. His hands were tolerably full of his work, the fruit of which is not discarded to this day. He had to take occasional lodgings in London to fulfil his secretarial duties. Hence it was that these duties were rather perfunctorily performed. The Foreign Correspondence was not being dealt with promptly and punctually (according to a Council minute of January 24, 1782).

A Committee appointed to consider his case (November, 1783) resolved that " for the benefit of the Society, the business of the Foreign Secretary be done by a person constantly resident in London." Shortly after this Dr. Hutton resigned his office.

On December 11 it was moved, at an ordinary meeting, that the thanks of the Society be given to Dr. Hutton for his services as Secretary for Foreign Correspondence. The President at first objected to this, on the ground that it was impossible for the Society, as a body, to know whether the duties of the Foreign Secretary had been efficiently performed or not. After a rather stormy discussion the motion was carried by thirty ayes against twenty-five noes ; and the President forthwith pronounced the Society's thanks to Dr. Hutton. The latter made a good defence in writing on the next evening meeting ; and it was resolved by a majority of thirty out of forty-five members present, that the Doctor had fully justified himself.

Matters might have ended here. But a storm had been gathering to which this episode was only preliminary. A strong party had been formed against Sir Joseph. Their ultimate aim was the choice of a new President in his stead. This feeling against Banks was precipitated by a sense, rightly or wrongly, that the mathematical side of the Society's work had been unduly shelved under the existing ruler. There were several very able men in the forefront of the opposition ; and, if to the above charge they could add that their President had improperly interfered in the election of Fellows, there were good reasons for questioning his suitability for the Chair.

Dr. Samuel Horsley (F.R.S., 1767) was a mathematician of some note, a well-informed man, and a fine preacher ; distinguished for his animosity toward the doctrines of Dr. Priestley and the warmth of his defence of orthodoxy generally. He rose to distinction in the Church, and made a respectable prelate. But his nature tended to partisanship : the very last quality to be desired in a President of the Royal Society. And it was understood that this was his ambition. Lord Brougham is far too contemptuous about Dr. Horsley's abilities ; but he is, perhaps, right in imparting credit to any persons who saved the Royal Society from " such a fate as being under Horsley's Presidency." Other vigorous opponents of Banks at this time were Francis Maseres (F.R.S., 1771), Cursitor Baron of the Exchequer, Political Reformer, and one of the acutest minds of the age ; and Dr. Nevil Maskelyne (F.R.S., 1764), Astronomer Royal, a profound mathematician, and an active member of the Society.

The conduct of Dr. Horsley and his supporters seems to have been hardly worthy of their position. It was their proper course to attack Sir Joseph Banks at a General Meeting, and not at the ordinary assemblies : interrupting the philosophical pursuits of the Society, and mixing up scientific with personal questions. The ordinary meetings of the Society continued to be marked with ill-humour, and with distinct hostility to the President. There was an organized party which did not disguise its determination to eject Sir Joseph from the Chair. Acting on the advice of a few friends assembled at his house, Banks made a " call " on the entire body of the Society to attend at the next ordinary meeting, as it was " probable that questions will be agitated on which the opinion of the Society at large ought to be taken."

This proceeding was resented by the malcontents. But Banks's friends meant business. On the eventful evening (early in January, 1784) a motion proposed by Mr. Anguish,[1] and seconded by the Hon. Henry Cavendish, " that this Society do approve of Sir Joseph Banks as their President, and mean to support him in that office," was carried by one hundred and nineteen ayes against forty-two noes. This ought to have been decisive. But tempers were as high as ever, a circumstance not likely to be ameliorated while the outside world was enjoying this quarrel of the philosophers. The news-

[1] Accountant-General (F.R.S., 1766).

papers did not spare them ; and the publicity which attached to their unfortunate proceedings doubtless helped to irritate some of the more earnest members of the Society.

Dr. Blagden to Sir Joseph Banks.

"(*February* 11, 1784.)

" DEAR SIR,—Gossett goes against you for his connexion with Maty. Dr. Warren brings down Gunning and all the friends he can muster for you. There is a paper in the *Public Advertiser* to-day in which Hutton's business is much misstated. You should get it. Horsley seems to take advantage of your staying in the Chair to represent the indecency of your sitting there to receive incense ; but still it is best. They say that many things will be brought out from quarters entirely unsuspected by you. Dr. Brocklesby is mentioned as one of those who, though they attended the meeting, do not think themselves engaged to vote for you, but I know not on what foundation. Brand Hollis meant yesterday to go against you. Warren says that there will be a thundering majority on your side, Jodrell that the majority is doubtful. The latter is growing violent against you, and probably will speak."

At the ordinary meeting, February 12, a resolution was moved that Dr. Hutton be requested to resume office. This was negatived by a majority of thirty-eight. On the next meeting, the opposing parties came to close quarters. It was moved, " that it would be highly indecent and improper if the President of this Society should hereafter, either in the selection of candidates or upon any other occasion, endeavour to avail himself of his situation to influence the vote of any officer of the Society." This was negatived by one hundred and fifteen against twenty-seven. In disregard of this ominous result, a second motion was persisted in : " that although it does not appear to the Society that either the present, or any former President, hath availed himself of his situation to influence the votes of officers in the election of Fellows, it is yet necessary to declare that it would be highly indecent and improper, if the President of the Royal Society should hereafter, either in the election of candidates, or upon any other occasion, endeavour to avail himself of his situation to influence the vote of any officer of the Society." This was negatived by seventy-nine against twenty-three.

Scientific evenings were now resumed ; and it was hoped there would be forgetting and forgiving, and a renewal of philosophic zeal. But Mr. Maty contrived to put himself into a false position by introducing, a month later, one of the anonymous pamphlets which had appeared on the topic. The President held the pamphlet to be slanderous and insulting, and the meeting appearing to be with him, he refused any official acknowledgment or notice of the pamphlet on the part of the Society. In the course of the evening Mr. Maty resigned his Secretaryship.

One more difficulty faced the Society before peace was quite restored. There were two candidates for the chief Secretaryship in place of Mr. Maty. Dr. Horsley and his friends brought forward Dr. Hutton for the post. The President made it a personal matter to befriend the candidature of Dr. Blagden. Weld suggests that it was vital matter for his hold on the Presidency to identify himself with the promotion of Blagden.[1] Before the day of election Sir Joseph circulated a card to the Fellows to the effect that, " at his desire," Dr. Blagden had declared himself a candidate, and the President did not doubt that his election would be of advantage to the Society. Meanwhile, some influential members met, led by Cavendish, Dalrymple, Aubert, Wollaston, General

[1] *Op. cit.*, II, 165.

Roy, Dr. Heberden, etc., formed a committee with Blagden, and resolved " to prevent a few turbulent individuals from continuing to interrupt the peace of the Society " (Blagden to Banks, April 5, 1784). Their efforts were successful in discountenancing any further obstruction or disunion on personal matters. After the election, which gave one hundred and thirty-nine votes for Blagden and thirty-nine for Hutton, the meetings quickly resumed their old peaceful career.

Sir Joseph bore his victory without elation, and, on the anniversary meeting of the Society, he addressed the assembled Fellows in a manly and generous way, as follows :[1]

" From the appearance of our present Meeting, I will venture to foretell that our disputes are at an end ; that the gentlemen from whom I have had the misfortune to differ in opinion will abide by the decisions of the Society, which they have repeatedly taken, agree with me in a determination to throw a veil of oblivion over all past animosities, and unite once more in sincere efforts toward the advancement of the Society, the honour and reputation of which we have all equally pledged ourselves to support.

" But, enough of dissension : a word never more, I sincerely hope, to be heard within these walls ; dedicated as they are by a generous Monarch to the service of Science. Peace and harmony should ever be found within them, for under the influence of peace and harmony among those who profess to cultivate it Science can only flourish. Let us unite once more then, my friends, to fulfil the wise purposes of our liberal Patron and Benefactor ; and resume at the same time the prudent conduct of your predecessors, who for more than a century supported the honour of this Society unsullied, and have bequeathed it to us as they received it. They never

[1] Weld, II, 168.

failed to sacrifice such resentment as rose among them to the good of the general cause in which they felt themselves embarked ; for, although some individuals among them have heretofore indulged their feelings by appealing to the public when they imagined the welfare of the Body at large was in danger, they never once attempted to convert the meetings instituted for the advancement of knowledge into assemblies of debate and controversy."

There was a deal of pamphleteering on this absorbing topic, while the newspapers and magazines took their share in the conflict. The general public was amused ; and, while it lasted, the matter gave unwonted glee to the classes who thought they could afford to laugh at philosophers squabbling. Of these ephemeral publications that of Dr. Kippis, who appears to have here summed up the case, perhaps shows the best literary merit.[1] Of the final conclusion of the case, he observes : " The man who, for a course of years, and without diminution, preserves the affections of those friends who know him best, is not likely to have unpardonable faults of temper. It is possible that Sir Joseph Banks may have assumed a firm tone in the execution of his duty as President of the Society, and have been free in his rebukes, where he apprehended that there was any occasion for them."

A few years after these occurrences there was question of Bishop Horsley's election to some Club, and of Banks's possible black-ball. In reply to his correspondent, Banks says : " For me, even had I been injured, custom would have allowed me to forgive. But, in truth, I was not. I was seated more firmly in my Chair by the

---

[1] *Observations on the late contests in the Royal Society* (London, 1784). Dr. Andrew Kippis (F.R.S., 1771), was a very able dissenting pastor, and led an active and useful literary life. He is now best remembered by an attempt to found a complete *Biographica Britannica* ; and by an excellent account of the *Life and Voyages of Captain Cook* (London, 1788, and still reprinted).

Doctor's attempt to dispossess me ; and our controversy ended, as I told him it would, on the very first day that war was regularly commenced. 'You have raised a storm, my good Doctor (said I), and, trust me, I shall ride upon it.' It is now eight or nine years since he left me unmolested to crow upon the dunghill he so valiantly disputed with me, and has not in that period made his appearance in the Society's rooms three times. To bear malice, therefore, against him would be insufferable. Was it in my power to be present, I would certainly ballot in his favour."

A few letters will illustrate the policy of Sir Joseph in his post as President. In order to preserve the credit and the honour of the Royal Society, it was inevitable that sometimes a peremptory tone was necessary.

Arthur Lee to Sir Joseph Banks.

" NEW YORK, *December* 2, 1787.

" SIR,—I received a few days ago a letter signed Geo. Gilpin, requesting me, in the name of the Council of the Royal Society, to order payment of £28 12s. sterling, stated to be due for eleven years' contribution, as a member of that Body.

" I should not have failed discharging annually my subscription, as a domestic member, during these years, had I conceived myself competent to sustain that character. But I thought, and still think, that the Declaration of Independence, and the Revolution consequent upon it, extinguished in me the character of a British subject, with all its benefits and burthens. It appeared to me that, without being a subject, I could not continue a domestic member of your illustrious Society. For that reason, and not from wanting a full sense of the honour of that character, I considered and consider myself as no longer a fellow of the Royal Society. I have, etc."

G

Sir Joseph Banks to Mr. Arthur Lee.

" SIR,—On the receipt of your favour I immediately laid it before the Council of the Royal Society, who accepted your resignation ; and have by that means prevented you from incurring any further debt.

" But, how the Independence of America can discharge you from the prior debt, which is an annual contribution, for the regular payment of which you at your admission into the Society gave your bond, is a matter which I confess I do not understand ; as a personal obligation thus voluntarily entered into appears to me equally binding on the person who enters into it, of whatever nation he may be, or under whatever Government he may choose to place himself."

M. Antoine Laurent Jussieu, the distinguished French botanist, was candidate for admission to the Royal Society. He was one of the early Foreign members of the Linnean. For some reason or other, unknown beyond the Council table, he was rejected. The following letter of Banks, apparently in reply to one which embodied surprise, or perhaps reproach, is an excellent specimen of his manner of treating a disagreeable subject in a manly and courteous way.

Sir Joseph Banks to M. Jussieu.

" SOHO SQUARE, *June 29,* 1788.

" SIR,—It is an unpleasant thing to be refused admittance into a Society to which one has been presented as a Candidate ; but in your case I consider it as a matter which ought not in any degree to vex or trouble you. I, Sir, who have had the misfortune to be rejected by the Academy of Paris, am now a member of it. Why then may not the same thing happen to you ?

" Your literary character, I can assure you, never came into the contemplation of those who voted against you. That is highly respected here, as much as you could wish or expect. As far as I know, your enemies were guided by an antipathy to the doctrines of Mesmer ; which, especially since Dr. Franklin gave his opinion against them, is here very prevalent. And, good Sir, difference of opinion in matters of speculative nature have not the least influence on the respect which literary men owe to each other. I hope that, notwithstanding your disappointment, our correspondence will continue. Mine with my friends at Paris did not cease a moment on a similar occasion.

" The son of Dr. Hope, the late Professor of Botany, will be shortly with you at Paris. He brings you two plates which are intended for the new publication of *Hortus Kewensis*, coloured in the best style. They are intended as furniture, and pay compliments to the Queen ; who studies Botany intensely, and really reads with perseverance Elementary Books ; also Lady Tankerville, whose Lord has a very fine Botanic Garden, and who knows plants well and paints them exquisitely. Accept these as a mark of my homage. Be assured of the continuation of that respect you have so justly inspired me with, and believe me at all times,

" Your most faithful," etc.

Sir Joseph Banks to Count Windischgratz.

" *June* 2 (?1784).

" SIR,—The Secretary of the Royal Society has orders to transmit to you the resolution of that Body, in which they decline the honour you intended them of assisting in adjudging the valuable prize you have offered in hopes to diminish that hitherto inceasing curse of civilization : the uncertainty of the Law.

" . . . Sensible of the benevolent intention of your question, and of the generosity which destined so large a premium to its answer, I regret that we are precluded from assisting in the decision upon it. But, as the Resolution which prevents the Royal Society from interfering on such occasions has been formed by the deliberate care of our predecessors, and has repeatedly been the means of maintaining that Independence which is our chief boast, I am fully satisfied of the great propriety of preserving it uninfringed.

" It may be necessary to explain to you, Sir, that the constitution of the Royal Society differs essentially from those of the Royal Academies of Paris and Berlin, though both of those academies appear to have been in imitation of ours as nearly as the policies of the respective Governments would allow. They are associations of learned men collected together by their respective monarchs, constantly called upon to answer such questions as their Governments think proper to put to them ; and held to the necessity of answering them whatever they may be, by pensions granted at the will of the monarch by whom the members are originally nominated. While we are a set of free Englishmen, elected by each other, and supported at our own expense, without accepting any pension or other emolument which can in any point of view subject us to receive orders or directions from any department of Government, be it ever so high.

" Witnesses of the French Academy having been called upon for decision on the most humiliating subjects, we have uniformly resisted when our own Government have called upon us for decisions. How evidently improper it would be for us, at the instance of a stranger, howsoever we may respect his rank and abilities, to do that which we have repeatedly refused to our own Government I leave to yourself to decide.

" That the refusal of the Royal Society will not (as you

justly observe) prevent your benevolent question from being judged, is a matter which gives me pleasure occupied even in the disagreeable task of conveying their refusal to you. All nations in Europe abound with Academies and Societies. Great Britain furnishes others beside us, and few indeed of those numerous Bodies will, I believe, be found to reject your offer, especially accompanied as it is with the annexed emolument.

" Accept therefore, Sir, my sincere wishes that you may find Judges worthy the benevolence of your intention, and that mankind may in future ages remember the benevolent author of it, with the gratitude that an act so universally public-spirited, and so worthy of one who really esteems himself a citizen of the world at large must be considered to deserve." [1]

Dr. Joseph Priestley to Sir Joseph Banks.

" 72 St. Paul's, *April* 25, 1790.

" Dear Sir,—As I wish always to act with openness and to avow the motives of my conduct, I cannot forbear to express my great dissatisfaction at the conduct of the Royal Society in the rejection of Mr. Cooper, recommended by myself and four other members, all men of science and of respectable character. There is not, I believe, another example of a certificate so signed and so slighted ; the votes, as I hear, being twenty-four against him, and twenty for him.

" My mortification is the greater, because it was in consequence of my own proposal that Mr. Cooper became a candidate. And, as I was known to interest myself in the business, by writing in his favour to both the Secre-

[1] Joseph Nikolas von Windischgratz. The Count took this refusal philosophically. In the following year, there appeared in London a Latin pamphlet, being an offer of prizes for improved forms of legal agreements ; under the title, *Ad Lectorem* : purporting to be translated from the German.—London, J. Nichols, 1785.

taries, and to my other friends in the Society, I consider
the proceeding as including in it an intended affront to
myself. Mr. Cooper, who was introduced to yourself,
and whose merit, independent of his certificate, was
attested by persons who have long known him, is a man
equally distinguished for his knowledge, ability, and
activity ; and of all the persons that I know, I think him
the most likely to do honour to any Society of which he
shall become a member.

" I consider the business as the effect of party-spirit,
political or religious, as highly unworthy of the Society,
injurious to the interests of philosophy, and arising
from principles which would equally lead to my own
exclusion from the Society. But, as I conceive it to be
a matter in which you, Sir, had no concern, it does not,
I assure you, diminish my respect for yourself, thinking, as
I have always professed to do, that the Society is honoured
by your being its President.

" I am," etc.

There may have been justice in the suspicion that
political partisanship had weighed some of the members
of the Royal Society on this occasion. Mr. Thomas
Cooper, of Manchester, was a man of learning and talent,
and love for natural science. But he had made himself
unduly ostentatious in sympathy with the French revolu-
tionists. Indeed, he was on the famous deputation of
English republicans who were honoured by a reception
from the national assembly.[1]

In those unpleasant times, a man of this stamp, however high his personal character, was simply an object of
fear. Whether rightly or wrongly, association with Mr.

---

[1] In 1793 he went to America, to see if it were " a place fit to live in."
Even in America, he suffered imprisonment on account of his advanced
opinions. Leaving politics alone, he presently became a judge, and
afterward a professor of chemistry.

Cooper, and others who thought with him, was shunned because of the supposed danger of such persons to the good of the commonwealth. Priestley, himself, was obnoxious to many. In writing this angry letter to the President, his tone is wholly partisan. The scientific work of Mr. Cooper is only mentioned by inference. There was none to mention, in point of fact ; considering the qualifications necessary for admission into the Society. Unless there existed tangible evidence of Mr. Cooper's scientific discoveries, it was certain that neither Banks, nor any other member except the candidate's personal sympathizers, could properly vote for him. So, this attack on the part of Dr. Priestley was a needless addition to the difficulties of the President's position ; at a time when friendships were shattered, and reputations endangered, by the merest whisper of party-spirit.

As Sir Joseph had nothing to hide, and nothing to excuse, his answer was easy. But he appears to have taken unusual pains to show his friend that his interference was uncalled for.

Sir Joseph Banks to Dr. Priestley.

" *April* 26, 1790.

" SIR,—In return for the openness of your conduct in your letter of yesterday, in which you tell me among other things, that you conceive I had no concern in the rejection of Mr. Cooper, I beg leave to meet you with the same candour by declaring that I am one of those who were not at the time of the Ballot sufficiently acquainted with the gentleman's merits to be justified in my own opinion in giving him my vote.

" At the same time, Sir, I assure you with the utmost sincerity that religious prejudice had no influence whatever on my conduct in that respect. For, though I am convinced that the majority governed have a right to insist that the magistrates who govern them do profess

the religious tenets which they believe to be the only
true ones, and conform to the rites by which the sincerity
of their religious profession can alone be put to the test,
I never felt the least difficulty in associating in philosophical disquisition with any person on account of what
his creed might or might not be.

" So much, Sir, for myself. On the part of the Royal
Society, whose Chair you do me the honour to say I have
not disgraced, I cannot help being astonished at the
sentiment in your letter, which expresses a disgust at
your recommendation to them being slighted ; and still
more at your conceiving the rejection of Mr. Cooper as an
intended affront to you. . . . To hint even, that the
Royal Society were capable of combining together in
order to pass an affront on you, is what they as a Body
must feel as a serious charge on their character and
reputation ; and is one, I boldly assert, which is void of
the slightest foundation in fact.

" With you, Sir, I have hitherto lived in friendship,
and have ever set, I trust, a proper value on the influence
your discoveries have had in the advancement of Science.
Your friendship I wish still to retain ; and I hope the
frankness of my letter will not alter the sentiments toward
me which are expected in you. I shall therefore in confidence add that, whatever Mr. Cooper's scientific merit
may be,—no token of which he has hitherto brought forward to the Society—there are, I am firmly convinced
reasons wholly independent of his religious opinions ;
which, however the partial eye of your friendship may
have overlooked them, do fully justify the late conduct
of the Society ; and will, if any appeal is made to the
public, acquit them wholly of the charges you are inclined
to bring."

This short misunderstanding with Dr. Priestley was
one of the minor difficulties which Sir Joseph had to face

in the exercise of his office as President. It required some delicacy to deal with each case as it arose. As it happens, there does not appear, in all his accessible correspondence, any example in which one decision conflicted with another. As he treated everybody with good humour, and an almost excessive politeness, it is likely that in few cases did he leave his correspondents offended. He had often to repeat that it was his duty, as was that of his predecessors, to examine with some degree of attention the pretensions of those who wished to belong to the Society, in points of view that might never occur to them nor to their friends, and he particularly seems to have set his face against " aspirants in the medical profession who could raise five guineas "; but who could not establish any record of original research.

In the same way, Sir Joseph had to dismiss a good many persons who offered ingenious papers for publication in the *Philosophical Transactions*, who had worked out discoveries for themselves, but did not know that Science was in advance of them. It was out of the question to submit such papers to the Society, seeing that its members were supposed to be conversant with opinions generally accepted.

M. Barthélemy Faujas de Saint-Fond made the tour of England and Scotland and the Western Islands, in the year 1784; a gentleman of some distinction, particularly versed in mineralogy and the infant prospects of Geological inquiry. He wrote a very agreeable book [1] on his experiences in this country; rather above the

---

[1] *Travels in England, Scotland, and the Hebrides, undertaken for the purpose of examining the State of the Arts, the Sciences, Natural History, and manners, in Great Britain* " (transl., 2 vols., London, 1799). A new and revised edition of this translation, with notes and memoirs, by Sir Archibald Geikie, has recently been published (Glasgow, 1907). Faujas is a man that should not be forgotten. He was one of those eccentric beings who believed in the possibility of scientific men not being jealous of each other, and was pleased to notice the generous tone in England of all whose company he frequented.

level of such productions, in the absence of useless sneer at matters which differed from the ways and customs of his own land. He was made a welcome guest in Soho Square ; and the fact that he dined with The Club one evening exemplifies the hospitality and the ready acceptance which were held out by Banks and his friends to European strangers of worth.

Banks's house was the " rendezvous of those who cultivate the sciences. They assemble every morning in one of the apartments of a numerous library, which consists entirely of books ön Natural History, and is the completest of its kind in existence. There all the journals and public papers, relative to the sciences, are to be found ; and there they communicate to each other such new discoveries, as they are informed of by their respective correspondents, or which are transmitted by the learned foreigners who visit London, and who are all admitted into this society. A friendly breakfast of tea or coffee supports that tone of ease and fraternity which ought universally to prevail among men of Science and letters."

Our traveller places on record the services specially rendered by Sir Joseph to the science of Botany : " he has become the guardian of several herbals executed by naturalists of great reputation. Had it not been for the attention and the fortune of Sir Joseph Banks, these collections would have been dispersed in distant quarters ; or, perhaps, lost by the negligence of heirs. Whereas, united as they now are in one repository, they are easily accessible to such as incline to consult them." What attracted him most of all, as being more in his own department, was the show of mineral products. Banks allowed him to select a few items : mostly varieties of Iceland spar, which he presently gave to the Paris Museum of Natural History.

The reception of Faujas at a dinner of the Royal Society

SPRING GROVE, IN HESTON, MIDDLESEX
Residence of Sir Joseph Banks

Club gave him great satisfaction. Sir Joseph presided ; and Dr. Maskelyne said grace. " To give liveliness to the scene, the President announced the health of the Prince of Wales ; this was his birthday. We then drank to the Elector Palatine, who was that day to be admitted a member of the Royal Society. The same compliment was paid to us foreigners, of whom there were five present."
. . . " Brandy, rum, and some other strong liquors, closed this philosophical banquet, which terminated at half-past seven, as there was to be a meeting of the Royal Society at eight o'clock. . . . I repaired to the Society along with Sir Joseph Banks, Mr. Cavendish, Dr. Maskelyne, Mr. Aubert, and Sir H. Englefield. We were all pretty much enlivened, but our gaiety was decorous."

Three days afterwards, M. Faujas was entertained at Banks's country house at Spring Grove (Isleworth). He was charmed with the management of the gardens, the state of their cultivation was admirable and interesting. An elegant dinner and dessert, "at which there was abundance of pine-apples," followed in the evening.

The day after this excursion, our traveller was entertained by William Herschel at his house near Windsor. Here he saw the " ever-memorable " telescope with which the eighth planet was discovered, and had the pleasure of making celestial observations during two hours. Herschel told him that he had made more than one hundred and forty mirrors with his own hands, before attaining the degree of perfection which he had now reached. Faujas was startled to find Miss Caroline Herschel assisting her brother in recording observations; " placed at the upper end of his telescope, when the indefatigable astronomer discovers in the most desert parts of the sky a nebula, or a star of the least magnitude, invisible to the naked eye, he informs his sister of it, by means of a string which communicates with the room where she sits ; upon the signal being given, the sister opens the window, and the brother asks her whatever information he wants."

Faujas was charmed with Kew Gardens. He properly gives the credit of their progress and utility to George III. Some of Banks's recent importations were on view. There were, especially, *Hedysarum gyrans* in flower ; *Dionæa muscipula; Magnolia grandiflora;* beside other plants which horticulturists of the period were more or less in ecstasies about. It is, perhaps, not generally known that the rockery in Kew Gardens was formed of lava fragments, which had been brought from Iceland as ballast. Doubtless, Faujas had this from Banks's own mouth. " As the lavas are full of cavities, fissures, and roughnesses, and are likewise spongy, and capable of imbibing and long retaining water, it was resolved to form thick borders of them, more or less elevated, round the verges of a shady piece of ground, appropriated to this moss-garden, which is unique of its kind."

Mr. Pennant, in a letter to Banks, tells of the visit of Faujas to the Isle of Staffa ; and of his fellow-philosopher hesitating on the shores of Mull, appalled at the aspect of Atlantic waves. The hesitation did not continue long. At much risk he carried out his plan, determined to follow in the footsteps of his predecessors, to whom he gives the credit of being the first to describe the wonderful island.

CHAPTER VI

KEW GARDENS—GEORGE III

AN important result from Banks's new acquaintance with the King was its effect upon the fortunes of the Royal Gardens at Kew. The care of the Gardens had been for some time past the favourite business of Lord Bute. Under his management the Princess Dowager of Wales was enabled to pursue her hobby in the best taste. Lord Bute was really one of the leading horticulturists of the time. In his younger married days, living at Mount Stuart, his time was occupied almost exclusively in Agriculture and Botany. The gardens there have been the glory of the island of Bute ever since. After his retirement from politics he resumed these things. At Luton Hoo he raised a splendid garden, from which he presently stocked another at his Hampshire villa near Christchurch. The general public knew nothing of the cultured side of Lord Bute's character. When an addition was made to the buildings at Kew he caused a private entrance to be made from his own study into the adjacent palace garden ; and in his hands it entered upon that career which has since made it famous throughout the world. The chief gardener was a somewhat remarkable man, William Aiton, a native of Lanarkshire. He came up to London to find employment, at the age of twenty-eight, which he found at the Chelsea Garden, under Philip Miller. His appointment as superintendent at Kew, in 1759, gave him his opportunity. Under his care the grounds flourished as they had never done before.

George III shared the affection of his mother for Kew palace and gardens, and fell in with her pursuits with zest, improving the premises and extending the plantations. Some of Sir William Chambers's buildings yet remain, in a style far removed from any fashion of our own days; but by no means inelegant, even in an age which has practically forsaken the seductive forms of palladian ornament. After the death of the Princess Dowager, the King bought the freehold of the house she had occupied. Some weeks later, when Banks came on the scene, and horticulture was found to have a common interest for the two, he fell into the position of Royal Adviser generally at Kew.

All that energy and intelligence could supply was now given to the improvement of the gardens. From the day when Banks had a tacit control of affairs, the acquisition of rare and curious flowers, and the discovery of useful Economic plants became his constant purpose. Without overshadowing the merits of the famous nurserymen of the day, or the praiseworthy efforts of those of the gentry who practised planting, Kew offered a perennial stimulus to emulation. His devotion to Kew was among the most signal benefits which Banks's career gave to his country.

A matter to which his attention was early directed was an expansion of the system of importing exotic plants. Hitherto acquisitions had depended upon the offerings of returned travellers, and upon occasional purchases from the professional nurserymen. By these means there was always scope for the gradual accumulation of rare and beautiful plants. From his own newly acquired knowledge of the strange and varied types in vegetation, so many of which might be transplanted to other lands, Banks conceived the notion of making Kew the depository of every known plant that could be of utility to the people of Europe, or ornamental to their dwellings. A recent foreign traveller in England has entitled Kew the *Mecca* of *Botanists*. No expression could so happily indicate the idea which gradually took possession of Banks, and which has so signally enriched the Science of Horticulture.

So it happened that, during a great part of the remainder of Banks's long life, an abiding interest was taken in the fortunes of Kew Gardens. After some thirty-five years or so of these activities, we shall come to the pathetic moment when, because of advancing age and infirmities, he is compelled to cease relations with a man whom he had trained as a Botanical Collector, and who now deserved the reward of years of diligence and zeal. Within this long period, it may be said there was a race of men come into being whose taste for natural science was actually inspired by Banks's renown. It was natural that exiled medical men, stationed in places far removed from the ordinary social amenities, should become ardent Botanists. It is more significant still that persons of all classes in life who were condemned to wander in foreign lands and seas, were aroused to habits of observation, and to take a share in the work of adding to the sum of human knowledge. An immense system of correspondence grew up, from persons in all parts of the world : so large, indeed, that it is not likely Banks was able to respond habitually. But the fact of the preservation of many hundreds of such letters is evidence that they were read and treasured ; and there is no lack of proof that the writers themselves were kept in mind. Indeed, Banks appears to have held all such persons in very close regard, always manifested without reserve when occasion brought them the chance of a personal interview. The letters from Banks's various correspondents reveal the astonishing activity which resulted from these botanical explorations.

Before many years passed away there was scarce any part of the world where there was not some one or other in touch with the now celebrated patron of Natural Science. A fresh interest was given to life. Whether at home or abroad, there was always some new devotee, inspired by Banks's generous example, seeking to unravel the long-neglected marvels of nature. During the reign of George III, nearly seven thousand new exotics were introduced into England. By far the greater part of these were sent home by Banks's plant collectors.

In connection with this enterprise, and partly fostered by the emulation aroused among botanists through its agency, there arose many Botanical Gardens in the British settlements abroad. The first of these was planted in Jamaica about the year 1775. In a very few years Jamaica became a planter's paradise, from the number of official and private gardens which flourished in the island. For there was everywhere a constant accession of zealous volunteers, both in the East and West Indies.

Francis Masson was the first of these adventurers sent out at the instance of Banks. As the first official collector sent out, and one whose services to Botany were not excelled by those of his colleagues and successors, his career may be specially noticed. He came from Aberdeen, and worked in Kew Gardens under Aiton for several years. He was about thirty-one years of age when Banks selected him for the pioneer Collector.

Masson went to the Cape of Good Hope, where he remained for three years. His first journey was more or less an experimental one ; yet his labours were rewarded by the discovery of many new plants, the consignment of which to Kew made a notable accession to the treasures in the Gardens. A second journey to the interior was made in company with Thunberg,[1] who had been sent out with a similar mission by the Dutch Government. This was a very fortunate circumstance for Masson, for Thunberg was not only a capable botanist, but companionable, and highly accomplished. They got as far as the Kaffir country, whence they were compelled to return from a successful botanic trip in consequence of the hostility of the natives.

In December, 1774, Masson sallied forth again upon another dangerous journey. He reached a point over five hundred miles northward from Cape Town, and brought away a superb collection of plants and shrubs. Four hundred new species of plants were consigned to Kew, and reached their new home safely. Among the more important of these were specimens of the beautiful *Ericas*, which have since flourished so well in this country under careful culture. (Indeed, it may be said that the Cape *Ericas* have greatly improved under European cultivation ; many of them were originally scrubby little plants whose chief merit was their novelty.)

In 1776, Masson was back in London. He was received with due honour. A paper of his was published in the *Philosophical Transactions*, LXVI. Banks was overjoyed at his success, and extolled everywhere Masson's zeal and intelligence.

Such was the beginning of a career that never flagged in usefulness. He travelled in Portugal, again in South Africa, and spent some years in the West Indies ; and

---

[1] Charles Peter Thunberg, afterwards Professor of Botany at Upsala University ; member of twenty-four European learned societies, including the Royal and the Linnean. He was in London in 1778 (*v. Resa uti Europa, Africa, Asia* . . . 1770–79), where he found his old fellow-student Dryander. He was taken to Banks's house, and received by Solander " in the politest manner." Banks showed him all possible attention. . . . " I accordingly spent the forenoon of every day in his house, and went all through his Herbarium, which was a most commodious as well as efficacious method of enlarging my stock of knowledge. . . . Several learned gentlemen assembled here as though it were to an Academy of Natural History."

H

at length went to Canada. Here his life ended prematurely, from exposure to the climate after so many
years in tropical and sub-tropical countries.

Banks remained a firm friend to Masson all his days.
He was proud of Masson's achievements, and was attracted by his personality. As he says, in a stray memorandum, Masson had ingratiated himself with all those
who made Natural History an amusement. And his
name is held in honour at Kew to this very day.[1]

King George III has sometimes been styled " Farmer
George." It is not clear how much, or how little, of
contempt is involved in this title ; but it is certain the
British public has never been enlightened on the King's
farming experiments.[2] Yet a great part of the King's

[1] *A Note on Plant Collecting :* It should be mentioned that this
obscure but useful profession was represented, before Banks's days, in
the persons of several remarkable adventurers. It would seem to be
almost a virgin field in literature to tell their story in all its rich and
varied fullness. There were enthusiastic plant-hunters from the days
of John Evelyn, by whose labours (as of hewers of wood and drawers of
water) the gardens of England were enriched with exotics. The whole
story of Sir Hans Sloane, for example, has never yet been told ; although
the materials are national property and quite accessible.

[2] There is, indeed, a portentous-looking biography with the title
Farmer George. But there is little or nothing of agriculture in it from
the first page to the last. We have not space here to make full amends
for this seeming neglect. But it is worth while, in addition to what
follows in the text, to present the following (unpublished) letter from
one of Banks's cronies.

Dr. Samuel Lysons to Sir Joseph Banks.
" CIRENCESTER, *August* 18, 1788.

" DEAR SIR,—You desired me to give you some account of what
His Majesty was doing in Gloucestershire. I shall be glad if I can tell
you anything you have not already seen in the newspapers. I spent
a few days at Cheltenham, and heard much of his conversation
on the walks. He appears much delighted with the environs of
that place, which are indeed very beautiful. He rode, walked,
and talked much, with the country gentlemen of that neighbourhood, who seemed much pleased with him, and he with them.
Mr. Hunt, the Chairman of our Quarter Sessions, who lives about a mile
from Cheltenham, seemed to be his most constant companion in his
rides. He experienced greater attention in the country than he has
received, I believe, for many years. It made Cheltenham very unpleasant as a public place ; everyone being expected to stand bare-
headed wherever His Majesty appeared. It is extremely diverting to

time was devoted to his farm. And his interests were shared with many enterprising country gentlemen. Sir Joseph Banks had become as ardent a farmer as he was a horticulturist. He seems to have carried the King along with him. About the year 1787, the question of improving the quality of wool was in the minds of most of the spirited country gentlemen who cared for their estates. The impulse given to the farming interest was largely

hear the various accounts the common people give of him. But they all agree in saying he looks like a very good-natured, merry gentleman ; those, indeed, who expected to see him with his crown and sceptre appear rather disappointed at finding that he wears a round hat and brown wig. He talks to the farmers whom he meets, and stay'd some time at a door to see some wool weighed ; and talked much about the price of it : His people (he said) fooled away his this year. Last Thursday he came into what we call the bottoms, to see the clothing manufacture, which was shewn him at Mr. Cooper's, a clothier who lives near Sir Geo. Paul's. The whole progress of the manufacture, from the picking the wool to the folding the cloth, was very judiciously disposed in separate divisions on the lawn of his garden. His Majesty seemed much delighted with every part of it, particularly the weaving and shearing the cloth ; and called the Queen and the Princesses about him to explain to them what had been explained to him : he told them they should wear scarlet cloth great-coats ; but they begged to be excused in the summer.

"It is said there were near twenty thousand people assembled round the place ; and I think it is possible, as that part of the country is very populous, and it was announced to every village within ten miles some days before. This visit was to have been on the day preceding ; and the following notice was cried by the bell-man of Minchinhampton, a town in the neighbourhood : ' This is to gee nautis that the King dwe not caal at Lord Geo. Paal's till Thusdy.'

"They breakfasted at Sir George's, who entertained His Majesty with the plans of the gaol and bridewell buildings in this county, and went from thence to Lord Ducie's. The country people were much delighted with hearing the King say it was one of the pleasantest days he had ever spent. He says that he intends coming to Cheltenham again next year, and reviewing the Gloucestershire Militia ; and he is then to visit the Duke of Beaufort, and Lords Berkeley, Bath, and Bristol. I fear I have tired you with so much on the same subject ; but we have, I think, nothing else to talk about here, unless it be to tell you the apples seem more numerous than the leaves in the orchards throughout the vale of Gloucester ; that the harvest is almost got in, even on our Cotswold Hills ; and the second crop of grass has been very abundant. . . .

"The Thames and Severn navigation goes on very well. They say it will be completed in a year. If the marriage of those rivers should be celebrated whilst the King is here, there will be fine doings, I dare say.

"With best compliments to Lady Banks and your sister," etc.

served by Arthur Young and Sir John Sinclair in co-operation with Banks.

Mr. Young is the most enthusiastic and industrious writer on agriculture ever known. His success in this line was due to a highly intelligent mind served by the pen of a ready writer. You find from Young that eloquence, once more, may be used in dealing with an economic topic. But these things could never have given him the great success in popular favour which he enjoyed, without a thoroughly practical manner of going to work. He led a strenuous life of experiment and observation. He made some mistakes and erroneous conclusions, and had to meet severe criticism ; but his extraordinary industry, and his unselfish public services, gave an authority to his utterances which could not be marred by occasional failure. In acquiring information, he pressed into his services every one, including the King himself, who was at all capable of adding to his stores of information. George III even put his pen to the service of the good cause.[1]

Sir John Sinclair was a man of different stamp from Arthur Young. Whereas the latter was kept comparatively poor, by his readiness for experiment, Sinclair had a wonderful power of influencing other men and inspiring them with his own public spirit. At his instance, many of the Scottish landlords, and some Englishmen, turned to agricultural improvement. The science of farming attained a certain amount of dignity. People began to see the value of true economic principles,

---

[1] *v. Annals of Agriculture*, 1786, 1799. Two papers signed Ralph Robinson, commended by the Editor as of " singular talent and clearness." They relate to the husbandry of William Duckett, a well-known farmer of the day ; who had invented a plough which turned well over the sward and threw it into each furrow, while the machine covered that up at the same time. Duckett rented large farms at Esher, at Ham, and at Petersham. He was awarded the gold medal of the Board of Agriculture and successive prizes from the Society of Arts. He died in 1802, aged 72. The King was associated with him for more than thirty years.

and their application to the duties appertaining to Property. Consequently other interests were improved : fisheries, house-building, clothing, sanitation, etc. were benefited by this awakening. Thus the two last decades of the eighteenth century saw the resources of the country immensely improved. Sinclair was a most laborious statistician ; without this aptitude, indeed, his plans and ideas could not have been so convincing. Because of all this, men like Sir Joseph Banks, whose own doings were not clouded in verbosity and pretence, recognized in Sir John Sinclair a true helper and reformer.

Sir John has left on record an appreciation of Banks, with especial reference to the fact that he did not meddle with politics. The latter wished to keep Agriculture out of that sphere, content that the Royal Society should continue to regard it as a scientific matter rather than an economic one. For this reason, Banks was at first opposed to the establishment of a Board of Agriculture. After, however, Sinclair had caused the President of the Royal Society to have a seat on the Board *ex officio*, and he had been induced to take some active part in its proceedings, Banks came to see the Board in a more favourable light, and eventually turned out a very useful member.

The King was heart and soul in these things. And from this period till the day when he was finally incapacitated by illness, his farms were matter of constant attention. We have no space to go very fully into the subject, but the fact of Sir Joseph Banks being concerned will be some excuse for giving the reader a peep into an almost unknown side of the King's personal career. When the authoritative life of George III comes to be written, there will be surprising revelations in every direction. For the present, the incidents now to be mentioned will go far to show what a very lively share he had in the activities pertaining to the life of the humblest yeoman.

Sir Joseph Banks to the King.

(*August* 10, 1787.)

" Sir Joseph Banks having been unsuccessful on Tuesday last at Kew, when he attended in the Botanic Garden in hopes of delivering the enclosed paper to Your Majesty, thought it best by last night's post to request that two Rams and four Ewes of the finest Spanish breed might be procured for him and sent home with all expedition.

" Knowing that the migrating flocks leave Old Castile, which is the neighbourhood of Bilbao, in autumn, on their journey to the confines of Portugal, where they spend the winter, he conceived the utmost despatch to be necessary lest they should have begun their journey before the order could be executed. Fearing also that whenever Dr. D——'s pamphlets, which Your Majesty has perused, fell into the hands of any well-informed Spaniard, the prejudice of Spanish wool being subject to degeneration which may have been raised by the French for the purpose of obtaining the Sheep would be done away, he is not sure but that he should have ordered a large number. But he can by this night's post either countermand those already ordered, if His Majesty's pleasure, which he will be ever solicitous to obey, is signified to him."

The King to Sir Joseph Banks.[1]

" St. James's, *August* 10, 1787.

" The King is much hurt he was not apprized on Tuesday last that Sir Joseph Banks was at Kew, (indeed, he never heard of it till he received his note this day) as he would have found time to have seen him. The King is much pleased that two Rams and four Ewes are sent for,

---

[1] The original of this letter is in the collection of Mr. A. M. Broadley.

and should wish the commission could be extended to twenty Ewes and ten Rams ; as, from the judicious remark of Sir Joseph Banks that Spain may soon find the evil of granting such Exportations, it may not be possible long to continue acquiring those useful animals.

" The King trusts this number from Bilbao will not stop the attempts of getting some through France, as well as others through Portugal.

" The extract of the letter from Madrid is not returned ; as it is supposed Sir Joseph Banks does not want it."

The King to Sir Joseph Banks.

" QUEEN'S HOUSE, *November* 29, 1787.

" The King is sorry to hear Sir Joseph Banks is still confined ; and though it is the common mode to congratulate persons on the first fit of the gout, he cannot join in so cruel an etiquette.

" He is glad to find that a small flock of sheep are near arriving, as he thinks it may be a means of improving the wool of this country, which he thinks a most national object. The hope that, through Portugal, Spanish Sheep will also be obtained, seems now to have a good appearance."

There must have been some activity at this time in the importation of Spanish sheep. This is not the consignment from Bilbao, recently mentioned (August 10); but a flock of some size which had doubtless been conducted by road from Spain through France, and transferred across from Calais. Mr. Walcot, agent for His Majesty's packet-boats at Dover, is in communication with Banks (December 26) on the matter of their embarkation. " . . . Captain Sutton of the *Union* packet-boat is particularly charged with this commission. I can venture to assure you, Sir, that a more attentive and able person could not have the execution of it. A barn, with turnips, coleseed, corn, etc. was bespoke, as well as a tilted cart if it shall be necessary. I have wrote to Mr. Mouron at Calais with the mail of this night. I wish, sir, to assure you that unnecessary expense shall not be incurred, though your desire of a proper liberality shall not be forgotten. . . ."

" Mr. R. Thompson at Dover has a very roomy good stable for the sheep on their arrival from Calais, and a field to air them in, with turnips, corn, etc. Also a steady Romney Marsh drover can be had to accompany Sir Joseph's servants with them to town, who is acquainted on the road, and knows how to get the usual drover's accommodation for flocks. Captain Sutton, if the weather should make it in the least necessary, will ' take every precaution to prevent the sheep from being thrown from side to side.' "

The flock was pretty large. But the affair was well managed. All arrived at Dover in very good condition ; one ewe having yeaned on board, and another dropped her lamb since. Others had family prospects. The number in all was forty-two sheep and six lambs ; and there was none of them lame, so the shepherd was prepared to start almost immediately for London. He counted on twelve miles a day.

Upon the first importation of the merino sheep (about 1790), the English and Scottish breeders were soon impressed by the extremely beautiful quality of their wool. The experimental admixture of breeds went on with considerable zest, and with much addition to the average weight of the fleece. It became almost the favourite subject of competition with the best noblemen and gentlemen of the day, whose first care in life was a high standard of condition on their own estates, combined with a belief in the need for nurturing the country's race of agriculturists.

The value of a good sheep of the new mixed breed
at length became extraordinary. The first ram sold at the
first sale of the King's flock, in 1804, fetched forty-two
guineas. The average price at the second sale was thirty-
eight pounds. Banks, writing to Mr. Dillwyn, tells of a
farmer whom he has just parted with who, getting a new
lease in Wiltshire, "meant to begin by buying a thirty-
eight pounder of the King's shepherd." In 1802 the prices
had been higher than this. Ewes were known to have
been sold at sixty-five guineas—Banks, on one occasion,
conveys an offer from a friend to give Lord Sherborne one
hundred pounds apiece for four rams.

In 1808, George III had a fine present made to him by
the Spanish Government, of a large flock of merino sheep.
They were 2214 in number, including about two hundred
intended for British Ministers and notables. These were
diminished in number by 427 that had died, either at sea,
or on the road between Portsmouth and Kew.

The flock was entrusted to Sir Joseph Banks for dis-
persal according to the King's wishes. They were not
to be sold ; but, as he stated to Sir John Sinclair, " they
are to be distributed to persons most likely in my opinion
to increase and improve them, at the charge of their
importation and expenses." His short circular to the
public runs thus :

" *Merino Sheep.*—All persons who possess ewes of
pure merino race, and are desirous of increasing their
stock, are requested to apply to Sir Joseph Banks,
who has received the King's commands to distribute a
considerable flock, newly imported from Spain, among
such persons as are most likely to preserve them free from
all admixture, and to improve their form by judiciously
matching them in breeding : giving a due preference to
those who have manifested their approbation of this kind
of stock by having already provided themselves with

the breed, but who have not yet obtained a sufficient increase to be able to supply the wants of their neighbours who wish to improve their British wools by the aid of this valuable cross. Letters addressed to Sir Joseph Banks, Soho Square, London, will be duly attended to. He requests to be correctly informed of the actual number of pure merino ewes, ewe tegs, and ewe lambs, each applicant is now in possession of, and of the source from whence the breed was originally procured.

" Sir Joseph Banks will be thankful to gentlemen who will inform him what was the average weight of the pure merino fleeces of the clips of 1808 and 1809, and what price per pound they were sold for, with the name of the purchaser."

In the spring of 1809 there was another immense deal in sheep which became of much interest to Banks and the leading breeders of the country.

Mr. Cochrane Johnstone wrote from Seville reporting that he had just bought " the finest flock of merino sheep that ever existed," twelve thousand in number, recently the property of a Spanish nobleman, an adherent of Bonaparte, and probably a sufferer in his cause. Both Sir John Sinclair and Sir Joseph Banks were communicated with. The difficult question of transport to England was the first ; a second one was how to deal with them when they arrived. He said to Banks, " This is a great national object, as great a one as driving the French out of Spain. . . . The King has to thank me for getting four thousand of them. I have got eight thousand by paying for them." A very good speculation if they could have been carefully got to this country and fed properly. As the matter stood, it was like presenting a schoolmaster with twelve thousand peg-tops.

Sir Joseph wrote to Lord Bathurst, asking what was to be done. Could he get an order from the Treasury for sufficient transports ? Bathurst was more inclined to discountenance the whole affair. Cochrane Johnstone had acted without authority ; and where would the Government be, if the consignees declined the responsibility which he had placed in their hands ? The sheep would be left on the hands of the Government at an enormous expense.

The consignees were Sir John Sinclair and Mr. William Cobbett : the latter, according to Cochrane Johnstone, " the friend of all thoroughly national objects, and as good a farmer as any in England." Cobbett, with his hands already full, took the public into his confidence, and announced the pending arrival of the sheep. He expects to have the care and management of them (out of friendship for the owner, unless he himself reaches London in time enough to look after them). He asks that any gentleman with some good wholesome pasture to spare for a month or two will be so good as to write him on the subject. He has himself provided plenty of pasture, but requires more, anywhere within twenty miles of Botley. The famous *Political Register* was simply boiling with controversial matter just now :[1] the War, Parliamentary Reform, Col. Wardle and Mr. Madocks, and the general Infamy of everybody. Yet Cobbett seemed, at such times, still an agriculturist first of all. It was so now. It would have been an odd thing (and not altogether unprofitable, perhaps), to see Banks and Cobbett as coadjutors in the disposal of this flock.

But it was not to be. We have no record of the final issue of this affair. As for Cochrane Johnstone, he was an adventurer pure and simple.

The probability is that a portion of this immense flock of sheep reached England, and that some fell to the King's share ; and, together with the lot of 1808, were being

[1] May–June, 1809.

dealt with by Banks. There are some allusions to the business in his available correspondence. We writes to his friend John Dillwyn in South Wales (September 18, 1809) :

" MY DEAR SIR,—It always gives me pleasure to obey your commands. In the present case, however, I have little hope of indulging myself. The flock of merino sheep, from which those were ordered for distribution by the King, were in the course of the campaign taken by the French, and soon after retaken by the Spaniards. The French, however, succeeded in driving off and securing 480 of the rams. The number, therefore, sent over and yet expected is very small indeed. It is not of so much consequence as it might seem, for it is quite impossible for me to proceed with the distribution this year (as the sheep already arrived are in a shocking state of feebleness) with any hope of doing right, till I can ascertain the numbers to be divided and the extent and number of the applications that will be made for them. I have already abundance, and every day's post brings me as many letters as half a day's work will answer."

In the summer of 1805 there was a misunderstanding between Banks and the King, the first in all their long friendship of more than thirty years. It began, of course, with a rather intemperate word on one side and wounded pique on the other. At the period in question, George III was beginning to display that hasty temper which was the prelude of his last illness. Sir Joseph was more ready to take offence than of old, and the utterance of a reproach on the part of the King wounded him to the quick. The circumstances are as follows :

Mr. R. F. Greville, in attendance on the King at Weymouth, told Banks by letter of His Majesty's general satisfaction at the success of the annual sale of Rams, but

was angry because Banks had " not kept faith with the
public," who were led to expect that the flock would be
sold by public auction only. Yet Mr. Coke had been
allowed to purchase some independently. Now Mr.
Snart (whose *status* as an official at Windsor is not easily
identifiable) had managed this part of the business
with Banks's concurrence, but without satisfactory
explanations to the King. In reply to Sir Joseph's
reprobation of the unkindly charge, Mr. Greville wrote
the next day to say that all was calm again. " The King
came up to me yesterday, and in great good-humour
said he now understood the business of the sheep better.
. . . I am writing to desire you to forget the flurry,
which I trust has passed away ; and you need not renew
the recollection by any marked explanation just now."

But Sir Joseph was not so easily mollified. He had
reason to believe that Snart had not dealt fairly with
him when reporting the business to the King. He sent
word to Snart that he was about to resign his connection
with the management of the King's flock, and wrote as
follows to Mr. Greville :

(*August* 20, 1805.)

" MY DEAR SIR,—I enclose a justification of my conduct
respecting the sheep, drawn up as concisely as I am able
to do it. I hope it will not appear to you too long.
Whether you use it in the whole or in part, and the
manner in which you introduce it, I must leave wholly
to your prudence and friendship. This Flurry, as you call
it, appears to me deeper than meets the eye. The ex-
plosion was delayed from the time you first mentioned
the circumstance till the Catalogue gave a proper opportunity for a vent. It was not the ebullition of a moment,
but a concerted menace, or I am mistaken.

" Great, therefore, as the misfortune will be to me
to lose His Majesty's good will, I must make up my mind
with fortitude to the event, whatever it may be ; and as

I have the consolation of being certain that I do not deserve it, I trust I shall be able to bear it.

" We have both observed that for some years past His Majesty's mind has been much more irritable and less placable than it formerly was. I do not now recollect an instance of any one, of whom he has said so hard a thing as he has said of me, being restored to confidential favour ; and it will be far better for me to be dismissed than to remain upon sufferance only. I feel a friendship for the King, and if it is returned as it used to be can never forego it. But coldness from a friend I can never support. . . ."

Greville replied with assurances that everything was most satisfactory with the King, who was " quite affected and much grieved that Banks took it so much to heart." But Banks told Greville frankly that he had not got over his surprise. He did not expect, if the King felt a friendship for him, to become the victim of anger originating in the want of information that might be had without difficulty. He did not like being dismissed by " a declaration that my intentions were good which implies that my conduct was bad." He pressed his wish to resign his duties. In the course of the following summer the King accepted his resignation, " if it must be," with unqualified approval of the zeal and ability with which Banks had conducted His Majesty's affairs.

But, after all this, we find Banks as full as ever of the King's farming business ; and he continued to be sufficiently busy with sheep until the terrible blow fell, which caused His Majesty's private affairs to be put into commission.

CHAPTER VII

PLANT COLLECTORS, ETC.

WHEN Banks, in May, 1771, was on his voyage home from the South Seas with Captain Cook, the *Endeavour* stopped at St. Helena ; and the naturalists made good use of their opportunity for studying the resources of the island. In the three days of their stay a great deal of useful information was gathered. Indeed, their active habits of observation must have had a strong impulse, as the reader will see who turns to the last chapter of Banks's *Journal*. There is scarcely a better account of St. Helena in existence.

One of Banks's acquaintances there was Daniel Corneille, an officer of the East India Company. This was one of the many men who were incited by his example. Early in 1787, Corneille returned to England for good, after twenty-three years' service with the Company ; the latter part of this period having been devoted to the study of Natural History. It was long before it occurred to him to communicate with Mr. Banks ; but as years rolled on, and rumours of Banks's renown which he would hear from passing ships reached his ears, he ventured upon claiming old acquaintanceship. He sent a dried fish, which he could not recognize ; and thought it might be worthy of a place with Banks's other natural curiosities.

Banks must have written a cordial reply to this letter. Corneille responded at some length, with an account of his proceedings, and his special endeavours in horticulture ; and an exhibition of his gratification on finding that his re-introduction was so welcome. He had been in communication with Masson, then at the Cape of Good Hope ; and suggested that he should be instructed to spend a few weeks at St. Helena. That Corneille was in earnest, and further encouraged by Sir Joseph's letters, is clear from the influence left behind him on the Island at his retirement. His successor was Robert Brooke, another officer of the East India Company ; a rather remarkable man, who had displayed good military talent, and a character for inflexible integrity during many years of active service. Since his retirement he had lived in Dublin, and taken pains to encourage the cotton manufacture in Ireland. His efforts were rewarded by what may be called moral success, accompanied by financial failure. On reminding John Company of his former services and present need, they made him Governor of St. Helena in succession to Mr. Corneille. He was active and sensible in his rather trying post, and originated several improvements. The legacy of Horticulture fell into excellent hands, and Brooke proved a good helper and correspondent of Sir Joseph. His gardener, Henry Porteous, seems to have been a practical and observant man ; coming well under the influence of Banks, and promising to become an accomplished botanist. Banks corresponded direct with him, apparently having discovered an apt pupil.

The East India Company carefully nursed St. Helena in those days. They wanted a capable Governor of the Island. After 1787 it was very advantageous to their service for European troops to be acclimatized here in preparation for India. According to Porteous, they had gardens allotted them, and milch cows : " As the Governor begged I would give them every assistance that lay in my power, in compliance with his orders I am always happy."

Colonel Robert Patton, who succeeded Mr. Brooke, about 1801, continued to promote the views of Sir Joseph Banks and to take a watchful interest in the St. Helena Gardens.

The East India Company had an appalling task at this period of their history. But their political troubles did not render them unmindful of the material benefits which it was possible to furnish for the people subject to their government. There was always a select body among the officials, high and low, that had these things uppermost in their minds. The great granary in Behar, raised by Warren Hastings, will always be an honourable monument to his memory. His successor, Lord Cornwallis, exerted himself in many ways to soften the rigours of existence among the native population. Cornwallis arrived out in 1786, as Governor of Bengal. At this date, plans were being arranged for transplanting fruit and other trees into Bengal which should improve the supplies of food. One scheme was to bring the sago palm from Malacca, in order to form a stock at Calcutta, from which a system of transportation to the principal towns in the province might be established, so that in time every village should possess the boon. The Persian date palm was another importation, which it was hoped would become a constant resource against famine.

Sir George Yonge was in frequent communication with Lord Cornwallis, and entered into these projects earnestly. It appears to be due to the pertinacity of Colonel Kyd,[1] that a Botanic Garden was at last established in Calcutta ; and it was presently placed under his superintendence.

Meanwhile, there were several ardent Naturalists, either settled in India or wandering about the archipelago. They were frequently in communication with

[1] Robert Kyd, Lieutenant-Colonel of the Bengal Infantry. Sir Joseph Hooker, who visited Calcutta Botanic Garden in his early days, has word of high praise for Colonel Kyd's work.

I

Sir Joseph Banks, or with Dr. Solander, who appear to
have held them in high esteem. Really, they should
deserve recognition with every one who knows the great
strides made by them in advancing Botanical Science.

Dr. König, for example, was a most remarkable man.
He was a native of Livonia ; had studied under Linnæus ;
and went on a botanic expedition to Iceland in 1765.
One of his discoveries there was the tiny plant since
called *Konigia*. He went to India about the year 1768,
as physician to the Danish settlement in the Carnatic.
His high qualifications were speedily recognized at all
the English and Danish settlements on the coast. His
simple and conciliating manners, and his unvarying
readiness to impart knowledge to others, together with
extraordinary enthusiasm when at work, captivated everybody. Having entered the service of the Nabob of Arcot,
as naturalist, he presently travelled ; [1] and afterwards
threw in his lot with the Honourable East India Company.
He regularly corresponded with Solander. He died in
1785, prematurely worn out with his exertions and
fatigues, leaving all his papers to Banks.

A very important character who figures in this section
of Banks's circle is Dr. William Roxburgh, the pioneer
exponent of the Indian Flora.

Born in Ayrshire in 1751, he became one of the more
distinguished botanical pupils of Dr. Hope. He started
in life as assistant-surgeon on an East Indiaman, and on
reaching Madras, in 1776, secured a similar appointment
at the General Hospital there. In 1781 he was stationed
as Surgeon to the garrison at Samulcotta. Here, in the
midst of his professional duties, he found plenty of scope
for the pursuit of his favourite science. His garden was
devoted to the cultivation of spices and Economic plants.
Coffee, cinnamon, the bread-fruit, the mulberry, etc.,

---

[1] *v. Journal of a voyage from India to Siam and Malacca, in 1779,* by
Dr. John Gerard König. MSS. in Nat. Hist. Museum.

were introduced into the plantation ; and he even tried
to produce silk, and to manufacture sugar. He soon
became remarkable for the devotion which he put into
his endeavours to improve the natural resources of the
country. Years afterward, we find him in charge of the
Botanic Garden at Calcutta, in succession to Colonel
Kyd. After a noble record of more than thirty years'
work in India he came home to England, and took high
rank with the veteran Botanists of the day. A fitting
monument to Roxburgh exists in his *Flora Indica*,
published after his death. While in India, that part of
his drawings, etc., which had survived the storms and
risks of transportation were collected by Sir Joseph
Banks, and published by the East India Company.[1]

Dr. Roxburgh first appears as a correspondent of
Sir Joseph in a letter dated March 8, 1779, when he
acknowledges a communication from him, and expresses
his surprise and delight at being found out : " About a
month ago I was honoured with your very agreeable
letter of March 25, 1778. Till then I did not flatter
myself that any collection of seeds or specimens of plants
I could make would be half so acceptable as you say, or
I would not have waited for your orders ; for I wish to
send such things to every person that will pay proper
attention to them, and not let them be lost." Then he
plunges into Botany ; and finishes with a reference to
Dr. König, who " is now on a voyage to the Islands in the
Straits of Malacca, and Siam. I hope to have my garden
greatly enriched by the labours of this voyage. He is
the most indefatigable man I ever saw. Many a day
we have spent together in the woods on this coast."
Roxburgh met Dr. König again at Tranquebar in 1782,
after his Siamese excursion, and henceforth much of their
botanical work was done in company. A hearty friend-

---

[1] *Plants of the Coast of Coromandel* (3 vols., folio), with preface and
notice of Roxburgh, by Dr. Patrick Russell.

ship was now grown up between them which lasted to the end. In December, 1785, he alludes to the death of König, whom all seem to concur in praising with considerable warmth. In his last will he left to Banks " all his manuscripts, and specimens of plants, with orders to me to send them by the first safe conveyance ; which I have now done by delivering them to the Governor of Madras with a request they may be sent home addressed to the Directors as part of the ship's packet." After 1790 Roxburgh's letters to Banks were very frequent. His indefatigable career in India lasted until 1804.

The successor of König to the post of Naturalist to the East India Company, in the Carnatic, was Patrick Russell, M.D. This was one of three clever brothers from Midlothian.

Alexander was settled a long time at Aleppo, whither Patrick followed him. They were active in studying and combating the plague. Alexander wrote a History of Aleppo, which was much esteemed at the time, while the leisure of his brother was devoted to Fishes, Snakes, and Plant-life. Claud, the youngest, went to India, appointed administrator of Vizagapatam. Patrick went out with him. From his very arrival, he began writing long letters to Sir Joseph Banks. Every six months an able and amusing chronicle poured forth from his pen. If it were possible we would include the whole of them in our story. As it is, a few extracts must satisfy us ; enough to display the merit of a good and kindly and witty man ; and an indefatigable Botanist who left enduring evidences of his scientific worth.

On his voyage out, Dr. Russell had leisure to finish the Medical History of Aleppo.

He reached Vizagapatam in December, 1782.

" . . . We landed at Tranquebar, June 6, and remained there till the 22nd. Dr. König arrived a few days before our departure, and I had the pleasure of making one short acquaintance with him. It was not safe, on account of Hyder's troops, to venture to any distance from the fort. He looked over my Aleppo catalogue, and informed me for my consolation that I must not expect to meet with above thirty of the plants in that catalogue in this country. I am in hopes of a visit from König at this place, as he talks of coming to the northward ; and my brother has given him a pressing invitation. In the meanwhile, I am not idle here. I have already collected a good many specimens, and purpose sending you at a venture such plants as I find marked in the *Hortus Malabaricus* [by Rheedes], or such as I am puzzled about in Rumphius [*Herbarium Amboinense*]. If you wish for any fresh specimens for your collection of better known plants, you need only send me a list, and I hope it is superfluous to assure you that I shall endeavour with great pleasure to procure them. . . .

" I unluckily left Willughby [1] behind me, and am totally unprovided with books on Birds and Fishes, except a small edition of Buffon which I found here ; and which, having few figures, is of little use for the Birds. I sometime ago directed Willughby to be sent me, but wish to have such books added as you may judge most proper on the subject of Birds and Fishes in this country. I likewise wish to make a small useful addition to my botanical library, which consists at present, beside the whole of Linnæus, the *Hortus Malabaricus*, and Rumphius's *Herb. Amboin.* I have wrote Solander on the subject, but request likewise your advice.

" We are at present in the most delicious season imaginable. Were it not for the agreeable circles round a good fire, it would chill all my faculties to think of spending another Christmas in London. England has one other attraction in skating ; and I am not without hopes of accompanying you to Hyde Park."

[1] *Ornithologia* (London, 1676; transl. by Ray, 1678).

(July 6, 1783.) " . . By the arrival of the last fleet, I learned with inexpressible concern, the death of my worthy friend Solander. I found I was not half the Stoic I thought myself."

(December 26, 1784.) Dr. Russell has news of the squabbles at the Royal Society : " Independent of other considerations, I am always pleased to see clerical pride mortified ; and though I have as large a portion of respect for Bishops as a worthy member of the Kirk ought to have, yet I have generally remarked, when certain objects are in view, that people of a certain denomination are less scrupulous in respect to the means employed in attaining them than men in other walks of life who affect less dependence on Heaven. . . . I have sent a small parcel addressed to you, with one for Dr. Hope ; and shall be very glad to hear how they thrive, for König complains bitterly that they seldom from Europe acquaint him with the success of the seeds he sends, which (he says) does very much discourage his spirits, and makes him lament sorely."

(July 9, 1785.) A very long letter mentions the death of Dr. König. From the details of his movements, it appears the Doctor set to work in an indifferent state of health : was too impatient to rejoin, and this in spite of Dr. Roxburgh having interdicted an immediate application to work. Two days before his death he sealed up carefully all the papers, etc., meant for Sir Joseph.

After a further complaint of want of books, Russell further descants humorously on the Dissensions : " In the *Monthly Review* for April, 1784, I found what was before a mystery to me cleared up. ' And dwells such bitter souls in holy men ! ' I find the first majority was 119 against 42, the second 85 against 47, the third 115 against 27, and then 102 against 23. ' Nevertheless the minority feel no abashment.' Some of them it may be presumed did feel abashment, if the minority dwindled from 42 to 23. The other half was made up of d—— impudent fellows incapable of blushing for anything. Had the texture of the skin been capable of erubescence, all their faces must have been as red as my Lord Kelly's nose, when their flatulent leader bounced with so little decency about secession. . . . Better not shew this at the *Shakespeare;* I do not mind Mr. Hamilton, but stand in awe of Pitcairne."

In November, 1785, the Presidency of Madras " politely nominated me as successor to Dr. König for the prosecution of researches in Natural History." This post he held for about three years, with satisfaction in all quarters. He sent a great deal of the results of his work to England. In reply to a request from the East India Company, Sir Joseph Banks wrote a lengthy and very favourable Report on Dr. Russell and his labours ; and readily offered his assistance in preparing their results for publication.

In further communications, he warmly praises his friend Roxburgh, who had lost (May, 1787) all his papers and most of his collections in a hurricane, yet had repaired his losses in course of time. He repeats his complaint of want of books. Of another, he says : " I once mentioned to you before that both König and Roxburgh complained of very seldom hearing of the success of the seeds sent home. I cannot help joining in their lamentation ; and whatever you may think of it, it is a real discouragement to us Wandering Collectors, of which I shall convince you at meeting." When, however, he was in London again (October, 1789), he was rewarded with hearty welcome, and appreciation of his splendid work. As for his Snakes, he found them cutting a respectable figure in the Museum, now neatly arranged.

"But I mean [he says] to re-examine them all." He presently published an *Account of Indian Serpents*. (London, 1796.)

It would fill many interesting volumes to recount the adventures of Banks's wandering friends. The above short detail will serve to indicate the vitality of all that was going on behind the scenes; while people at home were reaping the fruits of these arduous labours, and rejoicing in the sight of new and beautiful additions to their gardens.

The islands of Grenada, Dominica, St. Vincent, and Tobago having been ceded to Great Britain at the Peace of 1763, General Robert Melville was appointed His Majesty's Governor-in-Chief. He entered upon his duties on arrival at Grenada in December, 1764.

While making a tour of his Government, in the summer of 1765, the General reached St. Vincent's, and met there Dr. George Young, principal medical officer of the Island, and Surgeon to the Forces. Both of these men were examples of the public spirit which was a mark of the times. Melville suggested to the Doctor that now was the opportunity of establishing a Botanical Garden. A fit situation on suitable soil, with running water accessible, might easily be chosen for the purpose. The cultivation and improvement of many plants now growing wild, and the importation of others from similar climates, would be of great utility to the public, and vastly improve the resources of the Island. The Society of Arts made known by advertisement that they would give a premium[1] for the improvement of horticulture in the West Indies.

---

[1] The Society of Arts made a feature of their operations the giving of rewards of this kind. Between 1752 and 1776 they distributed either in medals or cash no less than £24,616. The Society filled a place in matters of practical economy, which the Royal Society did not undertake.

Doubtless, Sir Joseph Banks and Sir George Yonge were at the bottom of this proposal. Banks had been a member since 1764. General Melville put his heart into the matter. When the sale of Crown Lands shortly afterward took place, it was agreed that Young should apply for a piece of land for a Botanic Garden. An allotment was presently made by the Crown Commissioners, about a mile north of the town of Kingston. It was at first about twenty acres in extent. At his own expense, the Governor had it cleared and planted, and appointed Dr. Young superintendent. The ground was improved and cultivated ; and furnished with a great variety of rare and beautiful exotics, in addition to an ample culture of native plants. A dwelling-house and laboratory were erected. Here the Doctor remained in charge for a long time. In 1774 he reported progress to the Society of Arts ; and they rewarded him, in addition to their hearty approval, with a present of fifty guineas.

General Melville's governorship ceased in 1771. His successor was Valentine Morris, as Captain-General of the Island ; an old friend of Sir Joseph Banks, and an ardent horticulturist. He took up the interests of the garden in earnest. During the Anglo-Franco-American War St. Vincent suffered bitterly, but on the recession of the Island to Great Britain in 1784 everything revived. Alexander Anderson, who had been assistant-surgeon with Dr. Young, was made superintendent, and soon restored the garden to its former glories. He had many difficulties at starting ; but official interference was speedily nullified when Banks, and Sir George Yonge,[1] and other friends came forward and testified to his zeal and the value of his services. It appears from a memorandum of Sir Joseph, that the Marquis de Bouillé, French Governor of the Windward Islands during the war, met General Melville in London. They fraternized,

[1] Secretary for War, 1783–94.

as soldiers will in times of peace. On learning of the
partial ruin of Dr. Young's garden, and its confiscation,
de Bouillé insisted on its being restored by the new pro-
prietor. Anderson remained in charge of the garden.
The Society of Arts did him honour on account of
his services, in 1798, by granting him a silver medal,
and nominating him a corresponding member. When
Bryan Edwards reported on the St. Vincent garden
in January, 1792, Dr. Anderson was still on duty. "It
was a scene for a painter as well as a botanist," he
says.[1]

The success of Dr. Young's garden at St. Vincent gave
rise to the proposal of a similar institution at Jamaica.
There is a letter of the indefatigable Dr. Hope to Banks
(September, 1775), in which he announces the vote of the
Jamaica House of Assembly in favour of a Botanic
Garden. Ten years of loving attention to the work had
(in spite of some vicissitudes in consequence of the war)
borne abundant fruit. The Botanical Department had
been the means of introducing and propagating some
of the most valuable plants, especially from the East
Indies, which had now become staple products of the
Islands. Dr. Clarke, the nominee of Professor Hope,
arrived out in 1777, and brought with him a number of
exotics from England, including the camphor tree and
the sago palm. The mango was introduced by one of
Lord Rodney's captains, when that victorious officer was
at the West Indies in 1782. Beside this official or semi-
official work, several planters added their labours to the
enriching of the Island. The gardens of Mr. Hinton East,
of Kingston, appear to have been the greatest triumph
of the sort, in a private way. Matthew Wallen, another
settler of long standing, was an ardent importer of exotic
plants. His zeal went so far as to collect and acclimatize
the hardy plants of temperate climes. As Bryan Edwards

[1] *History of the West Indies*, 3rd ed., III, 262.

says,[1] " To Mr. Wallen the Island owes the water-cress, chickweed, wild pansy, groundsel, dead-nettle, dandelion, honey-suckle, clover, violet, the English oak ; his garden at Cold Spring is well-stocked with choice selections of introduced flowers, and European trees and shrubs." Mr. East, also, was not content without adding British plants to his garden. His list contains such items as the mullein, primrose, beet, carrot, celery, flax, asparagus, barberry, holly, elder, and London Pride.

These and other gentlemen were always in touch by letter with Sir Joseph Banks. The attention given to detail, and the zeal with which they urge extension and development, ran parallel with the warmth of their regard for the men who originated these schemes and who continued to lead public spirit in their direction so far away from home.

Hinton East was so successful with his plantations that, after his death, the Jamaica authorities bought his garden for the public use. He was specially desirous of adding to the island supply of Economic plants. In a letter to Banks (July, 1784) reporting progress, and offering novel suggestions, he remarks that the acquisition of the Bread-fruit would be " of infinite importance to the West India Islands, in affording a wholesome and pleasant food to our negroes, which would have the great advantage of being raised with infinitely less labour than the plantain, and not be subject to danger from excessively strong winds. The time is not very distant when measures will be taken by proper authority for bringing about this desirable event. . . . You have emboldened me, by your very obliging offers, to say that plants of any East Indian spices will be very acceptable."

The first suggestion of this project seems to come from Valentine Morris. In a letter to Sir Joseph Banks, dated St. Vincent's, April 17, 1772, Captain Morris states that

[1] *History of the West Indies*, 5th ed., I, 293.

he has considerable property in the Island, and that the population are not always too well-off for food. He proposes the introduction of the Bread-fruit tree, and is certain it would be the greatest blessing to the inhabitants.

Hinton East was in England in August, 1786. There is mention, in a casual note, of the *Senna Alexandria* (*Cassia acutifolia*), seeds of which Sir Joseph has given him, which will go to Jamaica the first opportunity ; where he fondly hopes " they may lay the foundation of an export of that article in due course of time to this country." A zealous man this Mr. East : one after Banks's own heart. Other letters of his, after his return to Jamaica, are in the same vein, always with practical and far-sighted views. But 'twas now easy, in personal conference with Sir Joseph, to raise with some energy the question of acclimatizing the Bread-fruit tree in the West Indies. When the idea was put before the King by Sir Joseph, it was adopted with some alacrity. It was just one of those schemes which appealed to the benevolent heart of George III. No long time elapsed before orders were given for getting ready a ship, on purpose for an expedition to the South Sea Islands.

Mr. Joshua Steele, of Barbados, was one of the increasing list of Banks's West Indian correspondents. As one of the early members of the Society of Arts, he belonged, doubtless, to Banks's circle in London. He went to Barbados in 1780, in order to manage his estates there personally. In this he was indefatigable, taking great care of his people, and setting a general example of well-doing. He was one of the foremost planters to grapple with the question of slavery ; and probably a leader of public opinion. In association with some other gentlemen, he started a Barbados Society for the Encouragement of Arts, etc. ; and proposed, in a letter to Banks (July, 1781), their being admitted, in an aggregate capacity, a member of the Royal Society in London. Banks was unable to promote this idea, as not quite in the scope of their operations. But a similar proposal was made to the Society of Arts, which was certainly a more appropriate plan ; and the Barbados Society was duly elected a " Corresponding Member."

All Mr. Steele's letters betoken a mind open to the welfare of mankind and the improvement of its conditions.

CHAPTER VIII

BLIGH'S VOYAGES

EARLY in 1787, it was determined that an Expedition be undertaken to the South Seas, with a view to collecting Bread-fruit trees for transplantation in the West Indies. Meanwhile, the idea had spread in the Islands, and more than one correspondent informed Banks that the French were determined to adopt a similar measure for their own possessions. As it happened, the British Government was forestalled in this experiment ; for a ship arrived at Martinique in the following year, with a cargo of Bread-fruit, cinnamon, and other plants in good order.

By the desire of Lord Sydney, Banks drew up instructions for David Nelson, the gardener. As a specimen of many similar documents, which Banks was always careful to write out for new collectors, it is introduced here in full. With its close detail, and abounding knowledge of whatever was necessary, it is a remarkable indication of his general ability to deal thoroughly with any matter he had on hand. And it is not without intrinsic interest as concerning the task in question.

" As the sole object of Government in chartering this vessel in our service at a very considerable expense is to furnish the West Indian Islands with the Bread-fruit and other valuable productions of the East, the master and crew of her must not think it a grievance to give up the best part of her accommodations for that purpose. The difficulty of carrying plants by sea is very great :

a small sprinkling of salt water, or of the salt dew which fills the air even in a moderate gale, will inevitably destroy them if not immediately washed off with fresh water. It is necessary therefore that the cabin be appropriated to the sole purpose of making a kind of greenhouse, and the key of it given to the custody of the gardener ; and that in case of cold weather in going round the Cape a stove be provided, by which it may be kept in a temperature equal to that of the intertropical countries.

" The fittest vessels for containing the plants that can easily be obtained I conceive to be casks, sawed down to a proper height, and properly pierced in their bottoms to let the water have a passage ; in both which articles the gardener's directions must be followed. Of such half-tubs, properly secured to the floor as near to each other as they can stand, a considerable number may find room in the cabin, each of which will hold several plants ; and these I consider as a stock which cannot be damaged or destroyed but by some extraordinary misfortune. As these tubs, which will be very heavy, must be frequently brought upon deck for the benefit of the sun, the crew must assist in moving them ; as indeed they must assist the gardener on all occasions in which he stands in need of their help. Beside these must be provided tubs so deep that the tops of the plants will not reach to their edges. These must be lashed all round the Quarter Deck, along the Boom, and in every place where room can possibly be found for them, and for each a cover of canvas must be made to fit it ; which covers it will be the duty of the gardener to put on and take off as he judges fitting ; and no one else must interfere with him in so doing on any account whatever,

" As the plants will frequently want to be washed, from the salt dampness which the sea air will deposit upon them, beside allowance of water a considerable provision must be made for that purpose ; but, as the vessel will have no cargo whatever but the plants on board, there will be abundant room for water-casks, of which she must be supplied with as large a quantity as possible, that the gardener may never be refused the quantity of water he may have occasion to demand.

" No Dogs, Cats, Monkeys, Parrots, Goats, or indeed any animals whatever must be allowed on board, except Hogs and Fowls for the Company's use ; and they must be carefully confined to their coops. Every precaution must be taken to prevent or destroy the Rats, as often as convenient. A boat with green boughs should be laid alongside with a gangway of green boughs laid from the hold to her, and a drum kept going below in the vessel for one or more nights ; and as poison will constantly be used to destroy them and cockroaches, the crew must not complain if some of them who may die in the ceiling make an unpleasant smell.

" As it is likely that the easterly winds will prevail to the south of the Line from the month of March to that of September, it is to be hoped that the vessel will be fitted out with as much despatch as is convenient, with a view of her not losing a year, which will be the case if she misses the first monsoon.

" Her first destination will be New Zealand, where she is to take on board two tubs of flax plants. From thence she is to proceed to the Society Isles, where she must stay till the gardener has produced a full stock of Bread-fruit trees ; and if Otaheite, which will probably be visited first, should not supply a sufficient number of such as are of a proper age for transplanting, she must proceed to Imao, Maitea, Huaheine, Ulietea, and Bolabola, and stay till enough are procured.

" She is next to proceed toward the Endeavour Straits, which separate New Holland from New Guinea ; and if she wants water in her passage she may put into the Friendly Isles in making the Straits, which lie in Lat.

10° 40′ The master must not be surprized if he falls in with a reef. He may be assured that with a little attention he may explore a passage through it. In these Straits he must find some harbour in which he may fill water, which there cannot be any difficulty in performing.

" From thence to Prince's Island in the Straits of Sunda will be the best run ; and if water should be wanted in the passage, it may be procured at Java, where the *Endeavour* watered. At Prince's Island the gardener will have some trees to get on board, which may make it necessary to spend some time there. From thence to the Isle of France will be an easy run, and from thence round the Cape ; at which place the ship must not touch unless there is absolute necessity. They must proceed to St. Helena, where she will receive orders from England pointing out the places in the West Indies at which she is to touch and deliver cargo."

The commander chosen for this Expedition was William Bligh, Lieutenant, R.N. ; who had already served in Polynesia under Captain Cook, as Master on board the *Resolution*. Since those days he had been with the Fleet in the West Indies ; and now reached home from Jamaica in August, 1787. He learned on arrival the " flattering news " that Sir Joseph Banks was awaiting him, having secured his appointment to the *Bounty*.

The vessel was fitting out at Deptford. Bligh at once joined her, and superintended the alterations in the cabin and elsewhere, according to Banks's instructions. The cabin was fitted with an extra floor, prepared for the reception of flower-pots, according to the wishes of the gardener, who preferred them to wooden casks. When the ship got round to Spithead, David Nelson saw to the embarkation of these pots, eight hundred in number, of various sizes, " made deep to make room for more shells for drainage."

K

Bligh was elated and cheerful during these preparations, as was very natural. That Sir Joseph had fixed his eye upon him during his absence could not but be very gratifying. Banks thought very highly of those men of Captain Cook's school who showed anything of promise. Several of them benefited in their profession through his personal knowledge of them. And a splendid chance had thus come to Bligh.

Lieutenant Bligh to Sir Joseph Banks.

"SPITHEAD, *November* 5, 1787.

" SIR,—I think the ship very capable. . . . The master [John Fryer] is a very good man, and gives me every satisfaction, and I think between this and the latitude of 60° S. I shall have them all in very good order. The conduct of Nelson and the garden is satisfactory, and we all seem embarked heartily in our cause, which I shall cherish as much as possible. . . . My Surgeon may be a very capable man, but his indolence and corpulency render him rather unfit for the voyage. I think it would be very proper for me to endeavour to get some young man as surgeon's mate, and enter him as A.B. . . . I trust nothing can prevent me from completing my voyage much to your satisfaction. Difficulties I laugh at. . . ."

A surgeon's mate was obtained for the ship, Thomas Ledward by name. Bligh's apprehensions over the surgeon were well-founded ; he died at Otaheite six weeks after arrival, from an illness aggravated by indolence and intemperance.

The *Bounty* sailed in November, 1787 ; but she was weather-bound in the Channel. A new start was made from Spithead on December 23, with a fine easterly breeze. Silence now closed over the ship's company for upwards of two years. At length a letter reached Sir Joseph Banks, dated Batavia, October 13, 1789, revealing the story of the mutiny, and the miserable failure of this well-intended and well-thought-out enterprise.

Lieutenant Bligh to Sir Joseph Banks.

" BATAVIA, *October* 13, 1789.

" DEAR SIR,—I am now so ill that it is with the utmost difficulty I can write to you ; but as I hope to be in England before you can receive it, the necessary information which may be omitted in this letter will be of no consequence. I have, however, for your satisfaction enclosed to you a short account of my voyage : it is nearly a copy of what I have given to the Governor of Coupang, and the Governor-General here, because my weak habit of body will not allow me to do more.

" You will now, Sir, with all your generous endeavours for the public good, see an unfortunate end to the undertaking ; and I feel very sensibly how you will receive the news of the failure of an Expedition that promised so much. The anxious and miserable hours I have past is beyond my description ; but while I have health the strange vicissitudes of human affairs can never affect me. Unfortunately, I have lost it at present, for, on my arrival here, I was seized with a fever, which fixing in my head made me almost distracted. But I am now better, and am to sail in the packet on Thursday next, which will save my life.

" You will find that the ship was taken from me in a most extraordinary manner, and I presume to say it could not have been done in any other way. I can, however, promise to you that my honour and character are without a blemish. And I shall appear before the Admiralty as soon as I can, that my conduct may be inquired into, and where I shall convince the world I stand as an officer despising mercy and forgiveness, if my conduct is at all blameable.

" Had I been accidentally appointed to the command
the loss of the ship would give me no material concern.
But when I reflect that it was through you, Sir, who
undertook to assert I was fully capable ; and the Eyes
of every one regarding the progress of the voyage, and
perhaps more with envy than delight, I cannot say but it
affects me considerably. To those, however, who may be
disposed to blame, let them see I had in fact completed
my undertaking. What man's situation could be so
peculiarly flattering as mine twelve hours before the loss
of the ship ? Everything was in the most perfect order,
and we were well stored with every necessary both for
service and health. By early attention to those particulars I acted against the power of chance, in case I could
not get through Endeavour Straits, as well as against any
accident that might befal me in them. And, to add to all
this, I had most successfully got all my plants in a most
flourishing and fine order. . . . Every person was in the
most perfect health, to establish which I had taken the
greatest pains, and bore a most anxious care through the
whole length of the voyage. I even rejected carrying
stock for my own use, throwing away the hen-coops and
every convenience. I roofed a place over the quarter-deck
and filled it with plants, which I looked at with delight
every day of my life.

" I can only conjecture that the Pirates (among whom
is poor Nelson's assistant) have ideally assured themselves
of a more happy life among the Otaheitans than they
could possibly have in England ; which, joined to some
female connections, has most likely been the leading cause
of the whole business. If I had been equipped with more
officers and marines, the piracy could never have happened.

" I arrived here on the 1st instant, and solicited the
Governor-General to be allowed a passage in the first
ship that sailed for Europe. But he has told me that he

could not possibly send us all in one ship ; and has consented, as granting me a favour, to be allowed to go in the packet ; for the Physician-General has represented my life in danger if I remain here. I am," etc.

Shortly told, the circumstances were these. The *Bounty* reached Otaheite on October 26, 1788, after a tedious and sometimes dangerous voyage. The first reception by the people was naturally doubtful, but Bligh's discretion, and the watchful eye he kept upon the dealings of his crew, soon won them over. The chiefs became hearty friends. They asked after *Captain Toote* and their former visitors. Everything was done to cement their good-will, and caution was exercised as to revealing prematurely their designs.

At length, the relations between the islanders and their visitors were altogether perfect. The process of lading the ship was begun, and pursued not only without difficulty, but on a basis of perfectly friendly trading intercourse. At the end of five months there were on board more than one thousand bread-fruit trees, beside many other plants. Everything was in Bligh's favour. So far his mission was accomplished. He sailed on April 4, 1789, with a " send-off " given by the chiefs to the ship's company worthy of the heartiest of such things at home.

But the *Bounty* had been too long at anchor. Six months of easy life in the tropics had demoralized the crew. Already symptoms of licence had appeared, in the desertion of three seamen, who carried with them the small cutter, with fire-arms and ammunition. On their being recovered, Bligh was absurdly lenient. They were abject in their expressions of regret, and their " steadfast resolution to behave better hereafter." With others, the resumption of discipline on board was irksome. There was nothing, however, in the conduct of any one

which might warn the Commander of impending disaster ;
and that probably happened only because of one master-
spirit, and his sudden resolve. Bligh was in good heart
with his prospects, and had no forebodings. There was
nothing in the conduct of any one on board which could
awaken any misgivings as to his due arrival at the West
Indies. Yet, early on the morning of April 28, the
Captain was awakened roughly from his sleep, bound and
thrust into the ship's launch which was already hoisted
out, together with eighteen members of his crew. This
daring outrage was the work of Fletcher Christian,
master's mate. He was twenty-four years of age only,
a gentleman's son with good education, and a promising
sailor. He had previously been on voyage in company
with Bligh. To the last, the two had been on quite proper
and apparently friendly terms.

The *Bounty* sailed away, to return to Otaheite, leaving
the launch adrift. Provisions, a compass, and a few
necessaries were thrown over to her occupants, now
absolutely helpless but for what energy and good fortune
might do for them. A memorandum book was luckily
among these things, so that Bligh was enabled to keep a
log. Bligh soon came to the determination to make
for the nearest Danish Settlement in the Indian archipelago, more than three thousand miles distant. After
undergoing great hardships, and losing one man at the
hands of savages, he made Timor on June 14. Here they
were very hospitably treated ; and after recruiting their
strength the survivors of the party reached Batavia,
and eventually embarked for Europe in October. The
exceptions were Ledward, the surgeon's mate, who is
supposed to have gone up country and settled there,
and five others who died from the results of fatigue and
exposure. One of these was David Nelson, the gardener.
Nelson's assistant, Brown, was one of the mutineers,
among whom were also the three wretches who had

deserted at Otaheite and been pardoned by their gentle-hearted commander.

Bligh landed at Portsmouth on March 14, 1790. The story of his heroic adventure across the seas had preceded him. Great indignation found utterance in England on learning the news of Bligh and his disaster. The Mutiny of the *Bounty* henceforth took a place among our most thrilling naval annals. In early Victorian days it was still a charm to work with boys ; and there may even now be found a few persons who know of it through the Story of Pitcairn Island.

Some months after Bligh's return home, Captain Edwards, in the *Pandora*, went in search of the culprits. The majority of them were long since gone from Otaheite, and all but one of those met with violent deaths. But several were taken, and brought away from Otaheite. One of these perished in the wreck of the *Pandora*,[1] the others were brought to court martial, and there were hanged. The fate of Fletcher Christian is uncertain. It is positively stated that he was seen in Plymouth streets, several years later, by one of his former shipmates. And there was a published account of his after-adventures, which, it must be said, however, bears all the appearance of being a *supercherie*.[2]

It was always, afterward, an open question as to the culpability of Captain Bligh for this misfortune, through faults of temper. There were officers of his own school who condemned him. This was, however, greatly due to the energetic advocacy of Edward Christian, a distinguished jurist of the period, who took his brother's part with much vehemence. But the matter is simple enough. No reliance can be placed on the oaths of the mutinous seamen

[1] *v. A voyage round the world in H.M. frigate Pandora, by George Hamilton, surgeon* (Berwick, 1793).
[2] *Voyages and Travels of Fletcher Christian, and a narrative of the Mutiny on board H.M. ship Bounty at Otaheite.* From the French (London, 1798).

who deposed to Bligh's violence. There is no reason to believe that he had exceeded the bounds of temperate discipline. Nothing was heard of complaint until the morning of the outrage. Really, the crew of the *Bounty* were demoralized by too long a stay among the nymphs of Otaheite. The resumption of discipline was too much for the weaker characters among them to bear. It was a simple matter when Christian, a fiery and energetic young man, subject to the like temptation to hanker after the joyous life so lately lost, whispered the seductive plan of ejecting the Captain and his officers and seizing upon the ship. What little was heard later of Christian is to the effect that he was a constant prey to remorse, and liable to exhibitions of ungovernable rage.

It is worth bearing in mind, that Sir Joseph Banks remained a staunch friend of Bligh as long as he lived. And this, although Bligh was presently the unfortunate victim of a very similar adventure, when his authority in New South Wales was shattered by the daring impudence of one subordinate.

This untoward affair kept the lawyers busy, as well as the whole naval service, for several years. The book and pamphlet trade likewise battened on it. Bligh was himself obliged to issue at least one rejoinder in paper-and-print. Among other things, he quoted a letter from John Hallett, a midshipman, who denied the ill-temper and bad language the mutineers attributed to him, in modest but forcible terms. He never heard " illiberal epithets " used toward Christian by his Commander, and always thought they were close friends until the morning of the catastrophe. At the court martial which presently ensued, Captain Bligh was honourably acquitted of the loss of his ship. He was presently introduced to the King ; and in course of time promoted.

The Jamaica House of Assembly showed their regard

for and sympathy with Bligh, by a vote of five hundred guineas.

We have no record of the language used by Sir Joseph when the news reached him of Bligh's disastrous failure in 1789. He must have been bitterly disappointed and angry. But he was not the man to sit down in idle resignation when a mischief could possibly be repaired. Scarce was the court martial disposed of, in October, 1790, than he began to appeal anew to the Ministry of the day. There is a note to Banks from Lord Auckland, dated the Hague, December 29, approving the proposal that another attempt be made to transplant the Bread-fruit tree. Sir George Yonge, soon after this, sent word to Banks that the King had been speaking to him about it ; and bid him confer with Lord Chatham. " I have only to add, the sooner the better. They have now ships of all sorts to spare. . . ."

There was little delay on this occasion. The *Providence* was fitted out, as the *Bounty* had been, with conveniences for plant transportation. The errand was again confided to Bligh. He was accompanied by Lieutenant Portlook and the *Assistance ;* and the two ships sailed for their destination in June, 1791. It was a long voyage. They made some fresh surveys and discoveries in Torres Straits ; called at St. Helena, whence Bligh wrote a good letter to Sir Joseph ; stayed once more at Otaheite, and eventually landed a cargo of plants in several of the West India Islands. Three hundred Bread-fruit trees were safely disembarked at Jamaica, and a like quantity at St. Vincent's.

It was now February, 1793. Hinton East was dead in the previous November, and did not, therefore, see the fulfilment of his wishes. He had been an ardent promoter of the first mission entrusted to Bligh. His anger was extreme when he heard of its failure ; and he expressed himself in very strong terms when commenting on it to Sir Joseph Banks. Many congratulations came from friends over the final success, beside public acknowledgments from the Jamaica Assembly and other authorities. Mr. Steele wrote from Barbados a long and enthusiastic letter, with news of the flourishing state of the gardens, and their encouragement upon the safe arrival of the new importations.

The botanists of this Expedition were Christopher Smith and James Wiles. The former returned home with Bligh. He was admitted to the Linnean Society (1793), and afterward went to Calcutta as Botanist to the East India Company, where he had a long and distinguished career. James Wiles elected to stay in the West Indies. He was placed in charge of the Bath Garden, at Jamaica. He wrote to Sir Joseph an interesting account of the progress of the plantations under his care. For years afterward he was in communication with Banks, entering with lively spirit into the wonderful progress of the exotic plants. The Bread-fruit tree was easily propagated by suckers, and was become common all over the island. The taste for planting, moreover, was on the increase. Several private gardens had been formed on a liberal plan. Wiles appears never to have returned to England. Quite happy in his congenial post, he remained there until his death in 1805.

The authorities at Jamaica could not but recognize the public spirit of Sir Joseph Banks. The House of Assembly passed a resolution of thanks to him " for his benevolent endeavours, exerted for the benefit of the West Indies in general and of this Island in particular."

CHAPTER IX

VARIOUS ADVENTURERS

Dr. Hope to Sir Joseph Banks.

"EDINBURGH, *August* 22, 1786.

"DEAR SIR,

I PRESUME to introduce to you the bearer of this, Mr. Archibald Menzies; who was early acquainted with the culture of plants, and acquired the principles of Botany by attending my lectures. He was particularly acquainted with the Scotch plants, of the rarest of which some years ago he made a collection for Doctors Fothergill and Pitcairn. He has been several years on the Halifax Station in His Majesty's service as a surgeon, where he has paid unremitting attention to his favourite study of Botany, and through the indulgence of the Commander-in-Chief had good opportunities afforded him. I am," etc.

This Mr. Menzies is one of the abler men to be brought into these pages as a Botanical Collector. He was from Perthshire, and was employed in the Edinburgh Botanic Garden while he studied for the Medical Profession. Dr. Hope is to be credited with a great deal of assistance in his education. Hope, indeed, had many " favourite " pupils during his Professorship, to the lasting honour of his College. Even Edinburgh has rarely had a more genial and painstaking teacher. He died in the autumn of this year, leaving a memory of sweetness and goodness in the breast of every one who knew him.

Menzies appears to have begun the serious work of life as assistant-surgeon in the Navy. We first hear of him in connection with Sir Joseph Banks, when on board the *Assistance*, at Halifax, N.S. From this place he sent botanical news, and a batch of seeds "for that noble collection at Kew." The *Assistance* reached Chatham in August, 1786. Menzies sent up to London, by hoy, a small box of Acadian plants, and promised a visit to Sir Joseph in a few days. Meanwhile: "I am informed there is a ship, a private adventurer, now fitting out at Deptford to go round the world. Should I be so happy as to be appointed Surgeon of her, it will at least gratify one of my greatest worldly ambitions; and afford one of the best opportunities of collecting seeds, and other objects of Natural History for you and the rest of my friends!"

Writing a few days later, he enclosed the above note from Dr. Hope, who, it may be, was unaware that Banks had some acquaintance with Menzies; or else, in the passionate courtesy of those days, was willing to cement the existing tie by a friendly interposition. With the same enclosure, Menzies announced that he was happily appointed Surgeon to an Expedition round the world. . . . Two vessels are going in company, a ship, the *Prince of Wales*, and a sloop, the *Princess Royal*. The proposed route is round South America, thence to the West Coast of North America, and by the Japanese islands to China, and then round the Cape of Good Hope homeward. Their chief object is the fur trade. But it is not allowed for the ship's company to trade or barter for any curiosities. "I hope, however, we are not debarred from picking them up when they come in our way." He asks Sir Joseph to intervene, if possible, with Mr. Etches.

R. C. Etches was the merchant, or ship-owner. Perhaps with an eye to good business, and doubtless impressed by the importance of conceding something to the influential P.R.S., he wrote to Sir Joseph, stating that he would dispense with the restrictions in the case of Menzies. Besides, he highly approved of the young man's conduct and manners. . . . " My younger brother is going the voyage, and I have given him orders to pay every attention to Mr. Menzies." The merchant further says that a gentleman has proposed to go as a passenger, with a servant, to Otaheite ; and to stay for two or three years. And he can accommodate another or two, if Sir Joseph knows of anybody. He could engage to fetch them back again in 1788 or 1789.

The *Prince of Wales* had a fortunate voyage round Cape Horn to the North Pacific Ocean. Menzies had just reason to be gratified when, nearly three years later, he reappeared in England. He had sent home a consignment of plants, and he had brought back the ship's company in perfect health ; only one man had died, and he from the consequences of intemperance. The following extract from his letter just before landing (dated July 14, 1789) must have been particularly interesting to Sir Joseph :

" On the west side of North America, in a remote corner inland, the natives had a short warlike weapon of solid brass, somewhat in the shape of a New Zealand *pata-patos*, about fifteen inches long. It had a short handle, with a round knob at the end ; and the blade was of an oval form, thick in the middle but becoming thinner toward the edges, and embellished on one side with an escutcheon, inscribing Jos. BANKS, Esq. The natives put a high value on it ; they would not part with it for considerable offers. The inscription, and escutcheonal embellishments, were nearly worn off by their great attentions in keeping it clean. . . . To commemorate this discovery I have given your name to a cluster of islands, round where we were then at anchor. In the course of a few days I shall have the honour of pointing out to you their situation and extent, on a chart which I have made of the coast ; as also of presenting you with a few mementos from that and other parts of it. Till which I am with due respect," etc.[1]

In these few years of peace, just preceding the outbreak of the French revolutionary wars, there was considerable activity in over-sea adventure. The French and the British Governments were emulous, alike with their own sailors, in discovering new lands and in seeking new outlets for trade and for colonial settlement. La Pérouse was away in the South Seas ; and the scientific world was beginning to be anxious concerning his fate. Our own traffic with Africa was increasing largely, especially at the Cape of Good Hope. New Holland [Australia] was now to be added to the lands familiarized with the sight of British sailors.

The year 1791 was remarkable for our maritime energy. Beside the regular traffic with Bengal and with Port Jackson, other important voyages were taken in hand. And, almost as a matter of course, Sir Joseph Banks was intimately concerned in them, even if he did not actually initiate them. One of these was a circumnavigation set on foot under the following circumstances :

Early in 1791, it was needful to come to an understanding with Spain about the possession of Nootka Sound. A projected voyage of discovery in the North Pacific had been postponed from various causes ; and it was now resolved by the Admiralty to combine the two objects in one Expedition, which should embrace a

---

[1] Nearly forty years later than this, Banks had a letter from the Rev. Thomas Kendall, of the Church Missionary Society, dated Bay of Islands [N.Z.], July 10, 1816, with a similar story : " Some time ago, being visited by some natives from the River Thames, one of them produced a brass maree, or war club, bearing ' Joseph Banks, Esq.' and your coat of arms engraved upon it. . . . The possessor would not consent to part with it for any consideration whatever."

survey of the western coasts of America hitherto un-known.

George Vancouver, one of Captain Cook's midshipmen, was selected to take charge of the Expedition. He had risen in the Naval service ; was a good all-round man, a strong disciplinarian, always careful of the conduct and the comfort of his crew, and specially apt as a nautical surveyor.

Archibald Menzies was appointed Naturalist to the *Discovery.* This was fulfilling his heart's desire. His tastes and his experience alike pointed him out as the man for the post. Later in the voyage he became Surgeon of the ship. The formal instructions of Sir Joseph Banks are unusually copious and interesting. They range over every conceivable topic which could attract the attention of a cultured and intelligent observer ; in Natural History, Botany, Social Science, and Physical Geography. His Collections and his Journals were to be considered the sole property of His Majesty. And, as many particulars would doubtless occur in the investigation of foreign countries that are not mentioned in these Instructions ; all such were left to his discretion and good sense : " You are hereby directed to act in them as you judge most likely to promote the interest of science and contribute to the increase of human knowledge."

There must have been open question as to Vancouver's temperament ; for he was instructed as to treating Menzies with proper distinction. And Banks's last letter to Menzies (August 10, 1791) plainly shows an apprehension as to their proper relations on board : " How Captain Vancouver will behave to you is more than I can guess, unless I was to judge by his conduct toward me,— which was not such as I am used to receive from persons in his situation. . . . As it would be highly imprudent in him to throw any obstacle in the way of your duty, I trust he will have too much good sense to obstruct it."

The *Discovery*, together with an armed tender, the *Chatham*, sailed in April, 1791. Vancouver visited the Sandwich Islands ; from thence went to Nootka Sound, where he established good relations with Spaniards ; returned along the coast of South America, visiting the Spanish Settlements and continuing the process of survey ; doubled Cape Horn, and anchored in the river Shannon, September 14, 1795. The whole voyage was passed without serious disaster of any kind. An excellent bill of health prevailed all the way.

Yet, this interesting yoyage was not destined to be without some unpleasantness at the last. There are always times when the kindest disciplinarian may seem to exceed the limits of his powers and of his own discretion. Thus it was with Vancouver. There is a story of a young midshipman, Lord Camelford, being flogged for some youthful indiscretion " at a gun in the cabin before all the officers." Menzies protests that the punishment was undeserved, besides being severe : and that Camelford will prove an ornament to his profession.

The relations between the Captain and the Surgeon were generally good. Menzies was permitted (according to Sir Joseph's wish) to build a glass house for his plants upon the quarter-deck. Yet, proper feeling could not endure when the Captain claimed possession of Menzies's journals, and the latter refused to give them up until Sir Joseph Banks and the Admiralty had granted permission. Presently, when Menzies complained that he had lost some of his best plants through his servant being taken off his duties, Vancouver put him under arrest for " insolence and contempt." This misunderstanding lasted from July 28 until their arrival in port.[1]

---

[1] By a memorandum of Banks (? Oct., 1796), it is plain that Vancouver was capricious and sometimes unjust to his middies. There are allusions to a published caricature of the fracas with Lord Camelford : the story, then, must have become public. Lady Camelford appears to have threatened action on behalf of her son.

Menzies did not lose his taste for nautical life. He presently served again in the Navy as Surgeon, and visited the West Indies. Afterward he settled in London, pursuing the practice of his profession ; and died at an advanced age in 1842. His Herbarium is in the Edinburgh Botanic Garden. His Journal was not published in full. The MS. was among those dispersed at the Banks sale in 1886.

Menzies is one of the names highly honoured at Kew. He was the discoverer of *Sequoia sempervirens*, of California. His name is immortalized in *Menziesia ferruginea*, a plant of North America allied to the *Ericaceæ*. The introduction of *Araucaria imbricata* was due to Menzies. The story goes that he was at a dinner given by the Viceroy of Chile to Captain Vancouver and his officers, where part of the dessert consisted of nuts which Menzies had never before seen. Instead of eating all his share, he took some with him on board, and having obtained a box of earth, planted them. They sprouted, and he succeeded in bringing five plants to England, which were safely received at Kew. One was planted in the Arboretum ; and another given to Sir Joseph Banks for his garden at Spring Grove.[1]

Among the philanthropic schemes of the period was an Association for Promoting the Discovery of the Inland Districts of Africa. The first committee of the Association consisted of Lord Rawdon, Bishop Watson of Llandaff, Andrew Stuart, Henry Beaufoy, and Sir Joseph Banks. There were ninety-five subscribers (at five guineas) ; but Beaufoy and Banks appear to have given liberal aid, in addition to unflagging personal attention to the carrying out of their plans. The hopes with which they started were not all fulfilled, nor were the actual results achieved anything like commensurate with the expenses incurred in treasure and in human life. Yet

---

[1] John Smith : *Hist. Records of Kew Gardens*, p. 287.

L

their failures may even be counted for victories, in the paths prepared for their successors. Without the pioneer work of the African Association, the opening up of the western parts of the Dark Continent might have been delayed for two or three generations.

The Association was soon at work. It happened that John Ledyard called one day upon Sir Joseph—Ledyard was from Connecticut, a genuine wanderer, and a brave fellow. He lived a long time among the North American Indians, worked his way to England before the mast, and joined Captain Cook's third circumnavigation, serving as a marine on board the *Resolution*. After this he was concerned in a journey to Kamtchatka, by land. Reaching St. Petersburg, he obtained twenty guineas from the Portuguese ambassador, on the credit of Sir Joseph Banks. At Yakutsk he fell in with Joseph Billings's Expedition,[1] and was prepared to join him. But he was suddenly arrested, taken back across Russia, and landed on the Polish frontier. Ledyard found himself at Königsberg, poor and ragged, and out of health. Yet somebody was found, in this remote quarter, who was willing to take his draft for five guineas on the President of the Royal Society !

He could now get to England. He soon waited on Sir Joseph, who was struck with Ledyard's manly but restless personality : a figure of great strength and activity, although but of middle size. His manners were easy, though unpolished, and he appeared to exhibit that rare

Billings was another adventurous spirit (? from Lincolnshire). He had sailed on the *Discovery* (sister-ship to the *Resolution*) as able seaman and astronomer's assistant. When William Coxe, the traveller, was in Russia in 1784, he recommended, through Dr. Pallas, to the Empress an expedition to complete the geographical knowledge of the more distant parts of the Empire. " Commodore Joseph Billings " was given the command of the undertaking, which occupied the years 1785–1794. Martin Sauer, Secretary to the Expedition, published in London (1802) an account of it and dedicated it to Sir Joseph Banks. Billings had been befriended by Banks more than once, and was most likely financed by him on this occasion.]

quality " of regarding all men as his equals which accompanies the man of supreme intellectual and physical energy." Banks received him very cordially, and told him about the proposals for opening up Africa. The very man that was wanted: he would be ready " tomorrow morning."

Ledyard left London June 30, 1788—barely a month having elapsed since the formation of the African Association. He was at Cairo in August, and spent some time there preparing for a route across to the river Niger. His first and only dispatch to the Association confirmed the Committee in the soundness of their choice. But their hopes were blasted by next having news that Ledyard had fallen victim to a bilious fever. He died at Cairo in January, 1789.

Meanwhile, another promising candidate appeared in Simon Lucas, sometime Oriental Interpreter to the British Court. He was sent to attack the unknown land from a north-easterly direction. He sailed for Tripoli in October, 1788, with a view to visiting Fezzan, passing the desert, and returning by way of Gambia. But a few days' forward journey compelled Lucas to abandon his plans. The hostility of the Arabs (he reported) was fatal to his hopes of making any progress. He remained at Tripoli as H.M. Consul for Morocco, sending to Banks and the other gentlemen occasional letters with information about the country

Major Daniel Houghton, 69th Regiment, was the third who volunteered to face the dangers of African exploration. With the experience of long service on the West Coast he had dreamed of the possibility of opening up the Niger country. Houghton reached the mouth of the Gambia river in November, 1800. The King of Barra recollected him and his kindly dealings, and gave him protection and assistance. All went well until he was at Medina, about nine hundred miles from the sea. Then disasters set in. He lost his trade goods, and his horse, and other property, including " the blue coat in which he had hoped to appear before the Sultan of Timbuctoo." Houghton's actual fate is unknown. He was reckoned a fine example of the British explorer.

In spite of these discouragements, the African Association persevered. They let it be known that they were prepared to give liberal remuneration to a qualified person who should offer his services.

Sir Joseph Banks was long acquainted with Mungo Park ; who returned from India, in 1793, whither he had been on voyage as assistant-surgeon to an East Indiaman. This appointment had been secured at the instance of Banks, who now recognized the maturing influence of his activities while absent from England. Park was in the flower of early manhood, vigorous in frame, and inured to tropical climates ; passionately fond of travelling ; an expert Naturalist. The offers of the African Association attracted him, and his services were readily accepted.

All the world knows of Park's intrepid first journey, which revealed so many of the secrets of the Gambia and hinterland. On his reappearance in London, after nearly three years' absence, the public were no less gratified than were the African Association at the issue of his wonderful exploit. Banks and his friends were exultant. They had, at last, a triumphant record for their pains. It was resolved by the Committee that Mr. Park had executed his commission with a degree of perseverance, industry, and ability that entitled him to the warmest approbation of the Society. They further insisted that Park's story be drawn up in narrative form and published for his own profit.

While Mungo Park was yet in Africa, the Committee sent out a young German, Frederik Hornemann, to whom Banks had previously given the opportunity of training. He learned Arabic, some natural history, and matters pertaining to his mission.   After an experimental trip, Hornemann left Tripoli in November, 1799, but was lost sight of for ever in the following year.

The African Committee now held their hands for a time. The inconvenience of occasional raids by the French on the west coast of Africa accounts, to some extent, for their inaction.   But, during the truce of Amiens, when a good many other projects were ready for exploitation which only a time of peace could justify, the Committee awakened to the reconsideration of plans for giving a Geography to their unknown Continent.   Sir Joseph Banks told Mungo Park that the mission to Africa was to be revived ; and, in the case of Government entering into the plan, Park would certainly be recommended as the proper person to be employed.

Ever since his return home, in 1797, Park had been dreaming of the subtle joys of travel and exploration. Beside a very large circle of appreciative friends in London, he had now a wife and a home in Peebles, and more good friends in Edinburgh.   But the quiet routine of a country surgeon's life was irksome to him.   His mind would get unsettled with expectations of being called upon to undertake fresh adventures in the desert and the jungle.   One such offer was actually made to him by the Government, in 1798: to go out with a surveying party to New Holland.   As far as can be gathered from his correspondence with Banks, the thing fell through on a question of the terms of his remuneration.   His intimacy with Sir Joseph appears to have been broken by this circumstance, who was a little offended at his fickleness as displayed in the negotiations.   It was not until the autumn of 1803 that matters revived ; when Park received a letter from the War Office asking him to attend in London.   After some little delay, and a pretence of consulting his friends, he accepted the Government proposal to take charge of a new expedition to Africa,

with a view to discover the outlet of the Niger. Writing
to Banks (October 20), he tells him of this offer of Lord
Hobart. He is to have a guard of twenty-five soldiers,
and a remuneration of ten shillings a day and £200
a year. The misunderstanding with Sir Joseph cleared
away like a summer cloud. Banks responded with warm
congratulations : " I have every reason to believe that
the plan, as well as the object of the undertaking, will
suit your opinions and ideas. As the offer made to you
seems handsome, I would by all means advise you to
return to London."

There was much to do in preparation. Park obtained
the assistance of a Moor, to reside with him at Peebles
while he learned Arabic. He improved himself diligently
in Astronomy and the use of instruments, and in the
methods for taking meteorological and magnetic observations. Thus thoroughly equipped, he finally left his
home in September, 1804, and reached London with a
prepared idea of his future operations. A memorandum
was given by him to the War Secretary, embodying
his plans. This document, of which the following is a
skeleton, is a remarkably clear exposition of the objects
in view ; and is a definite proof of his abilities, as it is of
the high spirit with which he undertook his task :

The extension of British Commerce and the enlarge-
ment of our Geographical Knowledge.

The investigation of the nature of the countries passed
through, especially with regard to their natural productions and the establishment of possible trade routes.

The articles of merchandise and their relative value,
and the extent to which the habits of the natives would
bear upon traffic.

The study of Natural History, and correct records of
latitude and longitude ; together with a survey of the
Niger, and of the Settlements thereon, and their inhabitants.

Ample lists of the men and animals required, their articles of dress and equipage; means for building two boats for use on the river Niger; and a list of various articles for merchandise and barter.

A brief programme of his intended movements, with suitable speculations as to the course of his route.

At the last moment, it was within the bounds of possibility that the enterprise would be relinquished. In conference with Major Rennell,[1] between whom and Mungo Park a warm friendship had arisen, the difficulties and the hazards attending upon it were so vividly shown, that Park was almost persuaded to withdraw from the undertaking. But when he was again alone, his enthusiasm revived. He found other persons in London who disapproved, but he faltered no longer. There was Sir Joseph Banks, fully as conscious as anybody of the dangers of the expedition, "one of the most hazardous ever undertaken," to remind him of the important objects in view, without trying to minimize the certain dangers which must be encountered.

On January 30, 1805, the *Crescent* sailed from Portsmouth having on board Park and his friends. These were Alexander Anderson, his brother-in-law, and George Scott, a draughtsman. The soldiers and bearers were to be supplied from the little garrison at Goree. In due time, the party reached Pisania, far up the river Gambia, whence Park had started ten years before. From this place he wrote (May 28) to Sir Joseph Banks with a short account of his movements; and to his wife, in very cheerful

---

[1] Major James Rennell was a close friend of Banks, and a lively correspondent. He returned to England in 1782, after a glorious period of service in the East; and at once became a distinguished member of the scientific world. Admiral Markham justly remarks of him that he " was the greatest geographer that Great Britain has yet produced." The maps of Africa, and especially those prepared for the illustration of Park's travels, were the first to put the geography of that continent into intelligible shape. He died in 1830.

vein. Not a single accident had occurred to mar their prospects. Then they departed, to face disaster after disaster. Sickness overtook the party. Many of them died; and when, in October, both Scott and Anderson perished, Mungo Park confesses to a gloom coming over his mind for the first time, producing an overwhelming sense of loneliness and friendlessness.

From Sansinding he wrote to Sir Joseph, and also to his wife; sadly, but not dreaming of the entire failure of the expedition. The final catastrophe occurred at an uncertain date; after the diminished party had passed Timbuctoo without being able to go into the city, and had gone some further distance along the Niger.

Another German, Roentgen by name, was also employed by the African Association. He essayed to reach Timbuctoo by caravan from Mogador. He was never heard of again. After him came another young volunteer, John Lewis Burckhardt, who had brought an introduction to Banks from Professor Blumenbach. Burckhardt sailed for Malta in March, 1809, and, after completing his Arabic training, completely explored Syria, the Nile Valley, and Bruce's track in Abyssinia. He was at Cairo in 1817, preparing for new conquests in travel in the direction of Fezzan and Morocco. But he was suddenly carried off by an attack of dysentery.

The work of the African Association was a matter of very great expense to the subscribers, as well as to His Majesty's Government. More good men perished in the swamps of Africa in their efforts to add to our geographical knowledge. After some years had passed, there was quite a numerous body of gentlemen who were making geography their hobby. The Raleigh Travellers' Club was the outcome of their association, consisting of such men as John Barrow, William Marsden, Basil Hall, Beechey, Marryat, etc. Presently, the African Association became merged in the new Royal Geographical

Society, and the Raleigh became the Geographical Club.

Another distinguished plant-collector was Adam Afzelius. Mr. Wilberforce having consulted Sir Joseph, on behalf of the Sierra Leone Company, as to the appointment of a Botanist in their service, conversant with tropical plants, Banks gave the name of his Swedish friend. Afzelius was in London in 1791, and welcomed the opportunity of going out to West Africa. The occasion added to his renown. He had the optimistic temper of his nation; and appears to have held himself always superior to the great trials and hazards of his occupation. From the first, Banks was careful to insist that his friend be treated with every consideration, and all proper provision made for him, as "a well-educated gentleman, and of high consideration in his own country." To this, Banks added his own friendly efforts to encourage Afzelius throughout the period of his exile.

Afzelius arrived outward at Freetown on May 6, 1792. He found the Colony in some disorder and confusion arising from the want of dwellings. But he put some energy into matters, and was able to get comfortable by the end of the year; and to send home a consignment of plants. His letters are always cheerful, until November, 1794, when he tells the story of a French invasion of the place. They simply sacked it. Everything was destroyed, and the settlers were left in a deplorable plight. "I saw my garden destroyed, my quadrupeds, birds, and lizards, all killed; my bottles containing quadrupeds, birds, amphibia, fishes, vermes, flowers and fruits, broke to pieces; my dry plants, fruits, seeds, shells, books and manuscripts scattered over the floor . . . trampled upon and covered with dirt, grease, molasses, rum, porter, bread, meat, bones, etc." To this was to be added the loss of his journal during the whole time of his residence;

microscopes, thermometers, barometer, and hydrometer. And the only thing to do was to sit down and wait until the arrival of a Company's ship, for the absolute necessaries of life.

This man was a veritable Mark Tapley. With all this trouble upon him, he was immediately at work repairing his losses in Natural History. It was several months before relief came for the settlement generally. But there is never a grumble with him. He only wants to repair his lost collections. When the Company sent out, in course of time, plenty of necessaries of every kind, but no instruments nor material for the prosecution of his work, he would have still been at a loss. But Banks's foresight had provided these things. And he writes to Afzelius, in his usual open-handed way, to the effect that he could draw upon him for any funds he wanted. By this letter (February 17, 1795) he advises him further that, seeing a Privateer may at any time come and renew the pillage, it will be better to finish his restorations and recoveries speedily, and return home by the first opportunity.

Sir Joseph had a slight misunderstanding with Mr. Wilberforce on his own account. The Sierra Leone Company found that Afzelius was sending bulbs and plants for Kew Gardens, and they claimed that his entire time should be devoted to their interests. Banks pointed out that he had supplied the Botanist with every article in his power likely to aid his researches, including a valuable collection of drawings. He had not put forward any claim on his own account, but such under the circumstances would not have been unreasonable. He added a suggestion that the Directors should send out plants themselves, according to the hint Afzelius had given ; the King would certainly spare some from Kew if it was necessary.

In 1796, Afzelius came home to London, with all his

collections. Thus ended a very useful career with the Sierra Leone Company. They had other vicissitudes to meet beside a sacking by the French. But they could not be entirely unfortunate with such a courageous and cheerful fellow as this on their staff. Afzelius became Secretary to the Swedish Embassy in London ; was elected to the Royal Society in 1798 ; joined the Linnean Society and contributed to its *Transactions*. In 1812 he returned to his native land, as Professor of *Materia Medica* in Upsala University.

CHAPTER X

MÆCENAS AND HIS HAPPENINGS

AS Sir Joseph Banks approached middle life, it was more and more obvious that he had unwittingly acquired a great hold upon the minds of his generation. The tradition of his wonderful reputation among his cotemporaries has been hitherto little understood. Clearly, he was a man of public spirit in the truest sense ; i.e. in rational work without palaver. Such a spirit was one of the needs of his time. He was by no means a solitary example of its exhibition. It was opportunely displayed in the persons of many eminent men whose activity and intelligence helped forward the progress of science and of philanthropic ideas.

It was Banks's happy fortune that he was not tempted, in his early years, to employ his ample fortune in a wasteful manner. Mature life found him the ardent devotee of science and the patron of every project of public utility. As his character developed with years, it was seen that a strong and sturdy man had arisen—a born leader. Beside this, he inspired much personal affection. His sagacious mind, his generous disposition, his independent habit, and his intimate knowledge of the world, readily captured the hearts of men. It is not surprising that the popular estimation of Sir Joseph Banks was so high.

Seeing the multiplicity and variety of his occupations, and the customary impulse to appeal to his opinion on

SIR JOSEPH BANKS
From a drawing by John Russell, R.A.

every possible novelty or difficulty, one is reminded of the legendary person quoted by the eulogist of King Solomon : " He was wiser than all men ; than Ethan the Ezrahite ; and Heman, and Chalcol, and Darda, the sons of Mahol ; and his fame was in all nations round about. . . . He spake of trees, from the cedar tree that is in Lebanon even unto the hyssop that springeth out of the wall : he spake also of beasts, and of fowl, and of creeping things, and of fishes."

This reputation for universal wisdom becomes a reality after minute examination of the Banks correspondence. Look, for example, at the following list of some few of the matters placed before him from time to time, each of them implying a special knowledge, together with a certainty in the mind of the querist that Sir Joseph was equal to the task placed before him :

All matters connected with the vegetable kingdom, as botanical novelties, horticulture, the cultivation of hemp, cotton, wheat, potatoes, apples, strawberries ; experiments in fertilization and the treatment of garden pests : these things were peculiarly Banks's concern. We find, besides, farming, drainage, docks, and canals ; hydraulics and engineering ; vital statistics ; tanning and currying ; " the tricks of millers and bakers " ; the plucking of geese ; taxation ; coinage ; earthquakes, magnetism and electricity ; geological gropings ; latitude and longitude. Exploration was a thoroughly live topic.

Some of these queries would be accompanied by samples or models ; as when an East India Company's official would submit a *vermiculus* or a *cimex* newly arrived in a parcel from the Indies. The contents of Banks's spacious cabinets were prodigiously augmented by these and other out-of-the-way offerings. Captains from the Eastern seas were entrusted with Japanese boxes, curious woods and minerals, wild beasts and birds alive or dead ; chests of tea " for Lady Banks " ; beside the inevitable bags of seeds, and collections of dried plants. Banks's friends all over the world believed in his capacity and resources perhaps more soundly than people about him at home ; seeing that he was, in a sense, their far-away lode-star. Their contributions were frequent and various, sometimes mightily curious, as " Indian cement for mending noses."

All these attentions were pleasing to Sir Joseph. His insatiable curiosity in the Natural Sciences had inspired an unusually wide circle of followers. The men who forgathered with him, alike with those who furthered his interests in foreign lands, were in a sense his disciples. Old and young alike rallied round him. The dignity of advancing years did not place under reserve that open, and kindly, and courteous way which had given a charm to his younger days and made him the soul of good company. At his table were grouped the surviving friends of his youth, together with young and fervent minds opening with the dawn of a new age. The cycle was never broken in Banks's lifetime : a period of ever-progressive culture, remarkable for its mastery over the secrets of Nature, and its pursuit of Philanthropic ends.

The numberless calls on Banks's time and attention would be of less moment to a man with fewer public responsibilities. But these he had in abundance. He had a surprising number of serious engagements, any one or two of which might well have employed a person of average versatility. Beside his absorbing functions in connection with the Royal Society, he was a diligent member of the Society of Arts, the Engineers' Society, the Dilettanti Society, the Society for the Improvement of Naval Architecture, and the Society of Antiquaries. All this involved regular convivial occasions ; for Banks was an ardent clubbist. As an

LADY BANKS
From a drawing by John Russell, R.A.

archæologist, he was second to none in his abounding
curiosity and zeal. As a mere matter of paper-and-print,
it would appear that there is better record of his doings in
the *Archæologia* than in all the volumes of the *Philosophical Transactions*. Sir Joseph sat on the Council of the
Society of Antiquaries from 1785 to 1787, and again from
1813 to 1820. Then, further, he was on the Board of
Longitude for many years, concerned in the superintendence of the Nautical Almanac, and similar abstruse
matters. Presently his services were required on the
subject of coinage, when he was made a Privy Councillor
in 1797. This honour was conferred upon Banks with
a view to his assistance in connection with the Board of
Trade, together with a committee on the new copper
coinage. This latter was a particularly laborious and
prolonged business. The first Lord Liverpool and Mr.
Matthew Boulton were long engaged with Sir Joseph on
this recondite subject; the former as the most learned
authority on coinage and currency, the latter as manu-
facturer. Sir John Sinclair and Arthur Young, with their
questions relative to the breeding of sheep, were seldom
absent from his mind; nor did he want occupation in
looking after the King's flocks and herds. The East
India Company frequently consulted him on Economic
products; and the Botanic Gardens at Calcutta, at
St. Helena, and the West Indies, had his constant
attention. Far from being overwhelmed with all these
things, Banks could take a part in the affairs of his parish;
and when at home in Lincolnshire, was as devoted to
county affairs as if he had no other irons in the fire
whatever.

Temporary concerns would arise sometimes, in which
Banks was expected to engage. He was usually considered an indispensable member when a good sound
Committee was to be formed. Such, for example, was
the case when the question of a monument to Samuel

Johnson hung fire for so long. Banks had been a pall-bearer at the Doctor's funeral in 1784. Several years were elapsed, and it was not yet decided that Johnson should have a monument in St. Paul's Cathedral. A strong Committee at length took it up, including Sir Joshua Reynolds, Windham, Burke, Boswell, Philip Metcalfe, Sir William Scott, and Sir Joseph Banks ; and the thing was settled.

Another interesting business was the revival of the Engineers' Society. Originally set on foot in 1771, under the presidency of John Smeaton, it presently suffered from disunion among the members, and ceased to exist after Smeaton's death, in 1792. In the following year it was successfully re-established on a better basis.[1] Sir Joseph Banks was one of the new "honorary amateur" members, without whose aid at first it is not to be expected that such societies can flourish. He purchased for the Society all Smeaton's papers and drawings and designs ; and the members agreed to undertake the printing at their own risk, pledging themselves to give the profits to Smeaton's family.

After the French war had begun, the usual inconveniences arose from stoppage of communications. This was a sore trouble to the scientific world, whose representatives on either side of the English Channel had a particular dislike to the interruption of their labours on the ground of international conflict. With such men as Banks, who knew nothing of politics, it was paralyzing. The requests made for the use of his influence in averting the incident troubles were cheerfully responded to. And it may be said that his taking action was generally respected.

There is one thing that Banks would never do. He would not interfere with politics or politicians. Not only because he was absorbed in his favourite pursuits

[1] Of sixty-five members in 1792, only fifteen were real engineers.

and in social improvements, but for the plain reason that he was conscious of knowing little or nothing of the points generally at issue. He kept away of set purpose from State matters. Cotemporary illusions as to the rapid perfectibility of mankind under French guidance he regarded with contempt—the contempt of ignorance. The only notice taken of persons given to social or political unsettling was a somewhat petulant refusal to have any dealings with them whatever. In truth, Banks was quite unfitted for any political association ; and he knew it. He uniformly rejected all overtures made to him to enter parliamentary life, though frequently and strongly tempted. Thus he saved himself a thousand vexations. Outside of party politics he could continue to be a personal friend of the King without incurring much displeasure from any one.

Banks's religious principles were as simple as were his political tenets. Just as he never troubled to understand the new thoughts on Government which were pervading the world in his days, so he never cared to take in question his own or any one else's creed. He was tolerant of all, but his was the tolerance of a man who felt that religious and political discussions did not concern him. He was contented with the " Church as by law established " (a familiar expression of the period), because he could be happy and contented under it, and saw that the majority of his fellow-countrymen were of the same mind with him. Of course, all his pursuits, from his earliest years, kept him necessarily apart from politics and polemics alike. Neither his rather narrow education nor his incapacity for transcendental inquiry allowed him to enter into such matters. But, as we have just said, he was tolerant. And he did not like intolerance in other people.

It is not often that religious topics occur in all his vast correspondence. Once in a while there is an attempt

M

to " draw " him on Biblical subjects : as when one asks
him about the mustard tree of St. Matthew, Banks
goes little further than to say that he has never studied
Biblical Botany, " not being at all confident that the
translators of the sacred writ have been as careful as they
ought in giving proper names to the plants mentioned in
different parts of the Holy Scriptures." Or, when an
old friend sends him a controversial pamphlet : "Sir
Joseph Banks thanks Dr. Shepherd, but freely declares
that he does not think a layman should spend much
of his time in studying St. John. He feels no doubt that
his Faith is sufficient, if his actions prove acceptable,
to conduct him to his home hereafter, believing that the
Strait Path pointed out by the other Evangelists will
lead him as safely to his object as the intricate one of
St. John, which Dr. Shepherd has so laudably attempted
to elucidate."

When, however, Banks heard of a case where some
word-splitting on these things was causing trouble,
breaking friendship, and destroying the public peace,
he was prepared with a strong opinion on the absurdity
of such quarrels. A good example occurs in the affair
of John Leslie, who succeeded Dr. John Robison as
Professor of Mathematics at Edinburgh. Leslie was a
friend of long standing (through the medium of Dr. Adam
Smith), and often had occasion to consult Banks upon
his affairs. There was a conflict of opinion when Leslie's
candidature appeared. It turned out that he had spoken
of an essay of David Hume as " a model of clear and
accurate reasoning." Certain ministers in Edinburgh
violently opposed the Town Council of the city for their
promotion of Leslie, because of his saying a good word
for the arch-heretic although purely in a literary sense.
Banks heard of this from two friends in London, and
unburthened his mind to Mr. Leslie on the whole
matter :

(*April* 19, 1805.)

" . . . I am sorry to learn that the clergy of Edinburgh, a set of men hitherto honoured for their mild and moral conduct as well as for the purity of their religious proceedings, have instituted what, in my humble opinion, much resembles a persecution of you ; for tenets which, if not strictly within their notions of orthodoxy, are surely such as wise and well-informed men should not select as proper objects of ecclesiastical censure. They would surely have acted more properly, and in a manner better becoming their station in the community, by suffering your book to remain quietly on the philosopher's shelf. . . .

" Surely a man may fulfil his duty to his Creator without assenting unconditionally to every undigested tenet which our half-informed predecessors have left behind as a legacy to their more enlightened posterity. If it has pleased God to permit his creatures to increase in wisdom, He will not condemn them for assenting to new opinions which their reason demonstrates to be just. . . ."

It is clear, from all this, that Banks was now enjoying a more than ordinary personal influence on his generation. It is, likewise, pretty certain that this was due to the development of a manly character in every sense of the word : quite as much as to his scientific position. There was another factor, however, which must not be forgotten in completing the estimate. He was notoriously wealthy ; and it was known everywhere that Sir Joseph was ready to spend his money in a good cause. Hence a large addition to his responsibilities in the shape of what may be called accidental claims upon his purse.

Strange and wonderful efforts were made by various sanguine persons to unloose those purse-strings. Few speculative undertakings, in Banks's active days, were conceived without the minds of the ingenious inventors turning, as the Needle to the Pole, toward the influential President of the Royal Society.

Banks was never averse to honest and wholesome enterprise. But he was not incautious, a circumstance which was unknown, or overlooked, in the first essay of a new-comer for his patronage. The temptations, there-fore, that were being continually thrown in his way— the schemes for which he was invited to share the expense —ranged over every possible flight. Novel inventions (with *perpetual motion* as a matter of course) ; exploring and colonization plans ; experiments in farming and horticulture ; proposals for reforming or correcting the existing state of things ; suggestions from persons desirous of pushing on in the world, or of retrieving a Past ; all followed one another with relentless certainty : all inspired by a determination to tap the resources of a wealthy and good-natured gentleman. So, with inventors and adventurers, curiosity hunters, would-be collectors of plants and birds, ship-owners who had in view new fields of venture, and personal friends who had suitors of their own whom they could not themselves gratify, together with appeals from men who had failed in life and were not likely to succeed because of their inherent weakness of character, Banks had upon his hands many and various interests which did not naturally belong to him, but on account of which he was called upon to pay.

Notwithstanding this plethora of other people's con-cerns, Sir Joseph usually met each new aspirant, or beggar, with geniality. But he spoke very plainly, especially when encouragement could not but be with-held. At times he would be really impressive ; as when an inventor either implied or enjoined secrecy concerning a new machine for raising water, or an improvement on the

camera, or other ingenious marvel—Banks would observe
that he had an insuperable objection to becoming the
depository of other men's secrets, lest he should bear the
blame of disclosing what other persons had not been able
to keep private. With all this, he did not like being
imposed upon. Neither was he willing to abet any
questionable attempts upon the resources of the Government. As he once said to a correspondent, " I find myself
so completely outnumbered by those who consider the
giving away of public money as a good-natured thing."

Banks's manner of refusal seldom appears to have been
downright repellent. Few persons had cause to hold
themselves finally disposed of. Some would return to
the attack after a decent interval of time. As the furtive
fly, misunderstanding the gentle action which has brushed
it away, comes back again to the feast, so these feeble
dwellers on providence persisted in hovering near the
generous fount of benevolence. An occasional applicant
of this sort, and a very sanguine one, was Henry Johnson.

Mr. Johnson propounded a method to Sir Joseph
(" with no other apology but the Voice of the World
declaring him to be the Encourager of Science and
Improvements ") for composing and printing, by entire
words and radices and terminations instead of single
letters as was hitherto the practice. Banks evidently
listened to these proposals, for Johnson presently writes
to say that his machine is set up, and can be seen in
operation. He hopes Sir Joseph will take the trouble
to come and see it. The invention was called *Logography*.
It appears to have been in actual use for some time. Sir
Joseph was one of the financial promoters ; and we find
the name of J. Walter, bookseller, Charing Cross, printer
of a newspaper, and founder of the *Times*, also among
them.[1]

The inventor was presently as poor as ever. Four years

[1] *v. An Introduction to Logography*, by H. Johnson (London, 1783).

later, he is asking Banks's assistance in his distress, accompanying his complaints with further ingenious notions. One of these is a sentimental Index to Shakespeare's Plays. Sir Joseph wrote him a kindly letter, with some straight advice, and an enclosure of five guineas : " You have my leave to place my name in that part of the list, whether first or anywhere else, that you think will do you the most service." Several years afterward, Johnson is again in want of help, and " cannot get employment for want of the expense of a few advertisements." He encloses a lengthy list and description of the schemes on hand. These include Indexes, Paper, Factitious Coals, White Lead, Acorns, Wood, Ashes, Bricks, Lottery Insurance. Logography, and Sierra Leone. Of the result of this application there is no record ; and the following, last of all this curious correspondence, likewise appears to have been left unanswered.

Henry Johnson to Sir Joseph Banks.

" *June 27,* 1792.

" SIR,—Hearing of your indisposition of the Gout, although not physically instructed, and hating a Quack in any profession, I cannot resist the impulse of informing you what has come to my knowledge in that regard."

Here follow two and a half pages 4to, chiefly founded on the late Dr. Allen's *Flowers of Sulphur, etc.*

" I hope, Sir, you will not attribute this information to officious impertinence, or commonplace prescription ; but to the true cause, a very anxious wish for the conservation of the health of a gentleman whose Indisposition at any time is so great a loss to Society."

Banks was very mindful of his old comrades in circumnavigation. It was always safe for any one of them who had won his personal regard to ask a favour of him. So it happened that several of Captain Cook's shipmates were on friendly terms with Sir Joseph all his life long. One of these, more important than others, was Sigismund Bacstrom, M.D., who had been taken on the *Adventure* for the second circumnavigation, and was withdrawn along with others of Banks's suite. He went to Iceland with the yachting party, and was in touch with Banks for several years. Afterwards he knocked about the world as surgeon on board various ships.[1]

Bacstrom renewed his acquaintance with Sir Joseph Banks in the year 1786. He was in distress in London, and remembered Banks's " kind and generous treatment " when in his service. He had since been on several voyages ; had given satisfaction and earned little. He hopes Sir Joseph will recommend him to a situation ; he could act as tutor to one or two young gentlemen, or attend an infirm gentleman going abroad for his health ; he knows several languages, and could assist in chemical experiments, "not the *Lapis Philosophorum seu potius Insanorum*, but honest investigation of the operations of Nature." Better than this, and more in Banks's line, he had discovered how to "increase the fertility of vegetation, by means of proper Natural Magnets." Corn, Wheat, Rye, and especially the Vine, received an enormous impulse from the use of his plan. Hearing that the Government intended colonizing Botany Bay, Bacstrom was willing to join the Expedition ; and would bring home plants, seeds, shells, minerals, etc. Some assistance was given to him in his immediate distress. We do not hear of him again until 1791, when once more Sir Joseph opened his purse, and wrote sympathetically. Matters now looked brighter. Bacstrom was fortunate

[1] He printed an interesting account of a voyage to Spitzbergen in the *Rising Sun*, Captain Souter (London, 1780).

enough to find other old friends, and regular employment. He must needs marry, thinking he was now settled in life. But he was tempted with another voyage round the world. The man's ingrained want of foresight, which was at the root of all his past troubles, again made itself manifest. The ship-owners undertook to provide for Bacstrom's wife during his absence. And he had to write to Sir Joseph for help in providing himself an outfit. Banks sent him £10, and encouraged him to collect plants, etc., for him.

Bacstrom was away on his voyages for nearly four years. The thing promised very well at first, and he went out in the best of spirits. At length, in November, 1796, more than five years after parting, Sir Joseph heard once again from him, with a wonderful tale of adventure and trouble. The Doctor had served in six different ships, had been taken prisoner three times, had been the victim of a mutiny, and had lost the valuable collection he had made for Banks of all sorts of curiosities. He had brought many sketches and drawings home, and had passed some time in pursuing a new system of philosophy. Grateful to learn that Banks had befriended his wife during his long absence, he craves the blessings of Heaven upon his benefactor's head. A subscription to Bacstrom's forthcoming work closes this rather sad history of a man who had certainly great abilities, but was handicapped with some obscure weakness of character which had hindered his whole walk in life.

Five years after this, Mr. T. H. Moseberg, architect and surveyor, wrote to Sir Joseph, with a proposal to colonize Surinam. His application is in Bacstrom's handwriting. There is no record of any result. It is pitiable to find a man of some natural parts hanging on to the acquaintance of Banks for upwards of a quarter of a century, unable to make substantial profit out of his opportunities.

Occasionally, Banks's prodigality in favour of the devotees of Science came home to him with some vexation. There is a curious state of affairs with the two Forsters, which reveals a determination on the part of Sir Joseph to insist upon his rights, however valueless the prospect of his money being ever repaid.

Johann Reinhold Forster came to England about the year 1766, in hopes of getting a living by Literature. With a growing young family, he found it difficult to make ends meet. His son [Johann] Georg was his ardent hope, and certainly showed much promise. Both father and son made friends among the learned, alike on account of their personal character and their mental gifts. They went as naturalists on the second circumnavigation. Socially, this arrangement was not a complete success, because of the father's irascible temper and frequent misunderstandings with Captain Cook. The result of their scientific labours was all that could be desired. But £4000 proved an inadequate payment for their services, after expenses were deducted. They were soon as impecunious as ever. J. R. Forster was glad to accept an offer from Banks of four hundred guineas for his drawings. After this, the Forsters proposed to go back to Germany. Georg insinuated the idea of a public subscription to enable them to do so. The father wrote to Banks (February, 1778) with a sad relation of the state of the family affairs. It must be about this time that Banks lent him £250. Then the two went to Germany, and succeeded better than they had done in England. Georg was a Professor at Cassel for several years, and corresponded freely with Sir Joseph ; on scientific topics generally, and sometimes on money matters.

J. R. Forster placed himself under the patronage of Frederick the Great. He became Professor of Natural History at Halle, and held the post with considerable distinction for twenty years. But he carried his financial troubles with him. He had to make an arrangement with his creditors. The Duke of Brunswick appears to have acted a friendly part, and probably aided the unlucky debtor. But, presently, it was discovered that within the schedule of his debts Forster had omitted to mention the £250 due to Banks. The latter was angry, and expostulated ; he would have at least some formal security, even if only the son at Cassel would give it. He did not expect to see the money again, but security he must have. His language to the agent of the Duke was very determined ; his letter rather warmly insisted that the legal proceedings in the case did not accord with his sense of justice ; and he meant to have it in his power to enforce the payment if ever Forster's circumstances should improve, " and no one can say that some time or other that may not be the case."

Georg Forster was not more fortunate in his pecuniary affairs. He was in arrears for five years' subscription to the Royal Society. Upon his request that they would excuse the debt altogether, Sir Joseph told him that the Council " were unanimous in opinion that his petition could not be granted, without a precedent likely to be disadvantageous to the finances of the Society."

In spite of these petty difficulties, Sir Joseph Banks was always kindly disposed toward the Forsters, admired their mental gifts, and was sufficiently desirous for their worldly advancement. The progress of scientific investigation was, after all, a passion with Banks ; this involved a sure regard for men like the Forsters, notwithstanding their inability to keep a good footing in their more worldly affairs. Nor was he niggardly toward them. J. R. Forster published at Halle, in 1781, in Latin and German, " Illustrations of Natural History, 15 plates, small folio. Engraved at the joint expense of Sir Joseph Banks, Mr. J. G. Loten, and Mr. Pennant."

Georg Forster held a Professorship at Wilna for some time, but returned to Cassel. His reputation never diminished during his lifetime, which closed prematurely in 1794. Among his numerous writings may be found a European Travel-book, containing a notice of the Fine Arts, etc., in England.

CHAPTER XI

THE SCOFFER ABROAD

THE foregoing chapter is unexpectedly interesting. The reason lies here : the material on which it is founded contains so many of the familiar aspects of elementary human nature. That magnetic attraction toward the great and the good, on the part of the less-fortunate and the half-hearted, may be said to belong to one of our First Principles. This remark is hazarded not in any spirit of contempt. On the contrary, if you give yourself time to ponder over the matter, it will be perceived that you are touching the very basis of civilization when you see men regard, with a longing, lingering glance, maybe—the pinnacles of success in life. There are persons of a shallow order of mind who will ascribe this attitude to the meaner motives. But this is an unkindly sentiment. To sound the praises of the victor, at Marathon or at Trafalgar, at Lord's or on Putney reach, is to raise the standard of your own potentialities. " When thou doest well unto thyself, men will speak good of thee," because the tribute of praise and honour is a secret and silent power for good works in the breast of every beholder. The sight of high principle in action is far and away the best educative force.

The reader needs hardly to be reminded that notoriety has its penalties as well as its gratifications. The hero of this short story, and his immediate friends and associates, were as lights shining in a mixed world. The

eighteenth century had certain social aspects greatly to
be deplored, upon which a vivid glare of light has been
thrown, casting into unmerited shade much that was
good. And the lovers of scandal have had their fill. No
one could be placed on a social pedestal, without being
exposed to the shafts of envious or jealous scribblers and
caricaturists. In the days of which we speak, slander and
caricature were at their best—or worst. A small tribe of
men, who either distrusted or despised the public recognition of virtue and worth, had arisen, that earned their
living by reckless defamation, and the best men suffered
with the worst.

The net result of these " lyttel gestes " is to be seen in
the character popularly ascribed to George III. Many
of the familiar fables (as, for example, the King's investigation into the making of an apple-dumpling) are trace-
able to Peter Pindar. Nor is Horace Walpole to be held
innocent of innuendoes which have served to distort most
lamentably the character of the King.

Horace Walpole had very friendly relations with
Banks. He was generally in touch with him on archæological matters, in association with Richard Payne Knight.
Marbles and vases, and costly editions of the Classics, and
the possession of unlimited means for the indulgence
of these tastes, cemented the association of these three
worthy men. Walpole's fancy did not go so far as Natural
History toys and playthings, but any new wonder in art
was sure to attract him ; and perhaps awake his ridicule,
as we know. Miss Berry met him at Banks's house one
evening :

" Where was a Parisian watchmaker, who produced
the smallest automaton that I suppose was ever created.
It was a rich snuff-box, not too large for a woman. On
opening the lid, an enamelled bird started up, sat on the
rim, turned round, fluttered its wings, and piped in a
delightful tone the notes of different birds, particularly

the jug-jug of the nighingale. It is the prettiest play-
thing you ever saw, the price tempting,—only £500.
That economist the P. of W. could not resist it, and
has bought one of these dicky birds." (*Journals*, etc.,
I, 286.) The few references to Banks which occur
in Walpole's correspondence are humorous, or contemptuous, as the reader likes to take them. As, for
example :

". . . this, however, is better than his going to draw
naked savages and be scalped, with that wild man Banks,
who is poaching in every ocean for the fry of little islands
that escaped the drag-net of Spain." (To Mann, September 20, 1772.)

" How I abominate Mr. Banks and Dr. Solander who
routed the poor Otaheitans out of the centre of the ocean,
and carried our abominable passions with them." (To
Rev. W. Cole, June 15, 1780.)

Horace Walpole is a representative Scoffer. But his
disinclination to take himself seriously is the worst thing
that can be said of him. It is more than probable that
he was one of those men who could be depended upon
for financial aid when useful or generous objects were
on foot. He had been a Fellow of the Royal Society since
the year 1746, so that he must have been familiar with
Banks's uniform public spirit. Evidently he had scant
sympathy with the Explorers. When James Bruce was
returned after ten years' absence, during which he had
contributed immensely to the sum of human knowledge,
Walpole announces his return home in short and sarcastic
terms :

". . . has lived three years in the Court of Abyssinia,
and breakfasted every morning with the maids of honour
on live oxen." (To Mann, July 10, 1774.)

This spirit was characteristic of the times. But it is greatly to be regretted that Walpole, who could produce workmanlike and trustworthy records of the past and many ingenious criticisms, seldom could write with dignity about his cotemporaries.

The Scoffer was abroad, everywhere. No one, in any stage of life, could escape him. Notoriety of any sort was exposed to his wiles. Lampooning was the rage; and clever lampooners could make money of it. The reader might have been spared needless allusion to this sickening topic. Yet it would hardly be doing justice to this interesting story were it altogether omitted; seeing that Banks, as has been shown, stood in the fiercest glare of publicity.

In point of fact, Banks proved an excellent target for the shafts of caricature. He was well-born, and hob-nobbed with sailors. He was a man of fortune, and reck-lessly defied the conventions in the disposal of his income. He was a personal friend of the King, a circumstance fatal in its relation to the discontented spirits of the day. His Presidency of the Royal Society was matter of ridi-cule with a small section of his colleagues, on the ground of his being no mathematician and certainly not another Isaac Newton. This last grievance, by the way, endured long after Banks's death.[1] And the little rift between the naturalists and the mathematicians was always liable to exposure.

The earliest caricature of Banks was harmless enough.

[1] The Rev. Thomas John Hussey (astronomer) sends to Charles Babbage, January, 1830, " half a dozen epigrams on a pair of Busts [in the Society's Hall], being a specimen of twelve dozen on the same subject." One " specimen " will be enough for us :—

" I think I've seen these things look very small,
I've seen a mouse in honest Cluny's stall,
I've seen a flea upon a lion's hide,
And Banks's bust with Newton's side by side."

(Addl. MSS., 37185/23.)

It was the day of the Macaronis,[1] who amused the town for a long season with published pictures of dandies and other eccentricities. Banks and Solander were, of course, lawful prey : with their renown in hitherto undiscovered trifles. The illustration opposite this page is a faithful reproduction of the honours paid to them on this occasion.

The *Fly-Catching Macaroni* is an etching, coloured by hand, representing Sir Joseph Banks " standing with one foot on each of two globes. The spheres are reversed to each other. The Antartick Circle of one, and the Artick Circle of the other, being uppermost. With a bat-shaped fly-catcher in each hand, Sir Joseph is endeavouring to catch a splendidly-coloured Butterfly. He has the ears of an ass, an ostrich plume is in his hat ; to his head is attached the Macaroni club of hair. Below the design these lines are engraved :

"I rove from pole to pole. You ask me why.
I tell you Truth, and catch a—Fly."

The *Simpling Macaroni* is an etched whole-length portrait of " Dr. Solander, F.R.S., standing in profile to our right, holding in one hand a large flowering plant, and in the other a naturalist's knife, on the blade of which is written the maker's name, ' Savigny.' He appears to be speaking. Below the design are engraved these lines :

"Like Soland Goose from frozen zone I wander,
On Shallow Banks grow fat, Sol * * * *."

After the publication of Cook's *First Voyage*, the world

---

[1] The term signifies " a compound dish made of vermicelli and other pastes, universally used in Italy. It came into England at the beginning of the last Peace " (*Macaroni and Theatrical Magazine*, October, 1772). The word was introduced at Almack's, and the subscribers came to be described as Macaronis. Originally aimed at luxury and extravagance, it eventually came to mean any person who exceeded the ordinary bounds of fashion and fell into absurdity in consequence. For more upon the Macaronis see *Annals of a Yorkshire House* (London, 1911).

THE SURRY MACARONI.

THE FORTUNATE MACARONI.

The SIMPLING MACARONI.

The FLY CATCHING MACARONI.

"MACARONIS." (CIRCA 1772)

knew more about Mr. Banks. It was clear that he had been the life and soul of the party, with his gun, and his dogs, and his sporting instincts. Cook is not chary in his admissions that Banks lent considerable spirit to the Expedition. Hence the newspapers found a new source of humour, in dwelling upon various incidents in which he was concerned. Especially seductive was the idea of his dallying with the tatooed ladies of Otaheite. Little of this is worth repeating at this distance of time. One skit in pamphlet form, rather clever, but desperately indecent, appeared as an Epistle from Oberea the Queen, by whom the visitors had been entertained and befriended.[1]

Some ten years after this period there appeared a new Light in lyric poetry, who kept the town going for upwards of a quarter of a century. He called himself " *Peter Pindar*,[2] a distant relation of the poet of Thebes,

[1] Attributed to Major John Scott, a very versatile pamphleteer of the day : *An Epistle from Oberea, Queen of Otaheite, to Joseph Banks, Esq. Translated by T. Q. Z., Esq., Professor of the Otaheite languages in Dublin, and of all the languages of the undiscovered islands in the South Sea ; and enriched with historical and explanatory notes* (4to, London, 1774). This was speedily followed by *An Epistle from Mr. Banks, Voyager, Monster Hunter, and Amoroso, to Oberea, Queen of Otaheite. Transfused by A. B. C., Esq., second professor of the Otaheite and of every other unknown Tongue* . . . (Printed at Batavia, for Jacopus Opano, and sold in London, 1774.) Opano, it will be remembered, was the phonetic form of Banks's name as pronounced by the islanders. An incident of Cook's second voyage is made use of by the unknown author of " The Travels of Hildebrand Bowman, Esquire, into Carnovirria, Taupiniera, Olfactaria, and Auditante, in New Zealand ; in the Island of Bonhommica, and in the powerful Kingdom of Luro-Volupto, on the Great Southern Continent. Written by Himself ; who went on shore in the *Adventure's* large cutter, at Queen Charlotte's Sound, New Zealand, the fatal 17th of December, 1773 ; and escaped being cut off, and devoured, with the rest of the boat's crew, by happening to be a-shooting in the woods ; where he was afterwards unfortunately left behind by the *Adventure*." (8vo, London, 1778.)

[2] i.e. John Wolcot, M.D., originally apprenticed to an apothecary at Fowey ; he found a patron who discerned in him some talent, and took him to Jamaica. Here he failed in making a physician's practice. He came home and took orders ! and returned to Jamaica, where he utterly failed as a parson. After his patron's death he returned to England, and started as a physician at Truro. Here he enjoyed but partial success on account of his general indiscretions. Far too witty a man for that provincial circle, he presently came to London, in company with young

N

and Laureate to the Academy." Beginning with the
painters of the day (1782), he assailed in turn every
public man from the King downward. Every bit of
gossip whatsoever was sure to be repeated in the version
of Peter Pindar. He seemed incapable of praising any
one. Unlike the gentle genuine humorist, he could
palliate nothing. Derision was the beginning, and derision was the end. The versification was marvellously
good, yet the eternal sneer reduced it to twaddle.

After the King, and the members of his Ministry, and
some of the Royal Academicians, Sir Joseph Banks was
Peter's favourite quarry.

> " High o'er the world Sir Joseph soars sublime,
> The great and fertile subject of my Rhyme," etc.

He is a peg from which hangs a selection of hints
which can serve for throwing in at any juncture.

> " Go to the fields, and gain a nation's Thanks,
> Catch Grasshoppers and Butterflies for Banks."

Generally, we learn that Banks made of the Royal
Society a Fly Club, and tried to amuse them with frogs,
and flies, and grasshoppers, and weed-and-birds'-nest-
hunting ; that he was overbearing, and kept the members awake by loud strokes of his official hammer ; that
he sometimes swore ; that he was " too common, too
ignorant, and too vain for the chair of Newton." He
was a lime-and-mortar Knight, whatever that may mean.
He proposed the plan of a throne for himself, " and
benches for foreign Princes and Ambassadors beneath

John Opie, whose merit he foresaw and resolved to encourage. Some
kind of partnership between them existed for a time. At length, finding
London would neither have him as a priest nor as a physician, he took to
his pen, and carried the town by storm, with " Lyric odes to the Royal
Academicians for 1782." His verse is by no means discreditable, except
for its daring personalities, and some indecencies which we could not
now endure.

T. Rowlandson

A FEAST AT THE FLY CLUB

him, whose heads might be on the same plane with the most noble President's ten toes " ; but

> " Th' uncourtly Doctor [Horsley], hostile to the scheme,
> Gave a loud *horse laugh*, and dissolved the dream."

When a little boy at school, he " munched spiders upon his bread-and-butter." Now : " his dinners are the wonder of the Nation."

Worse than all this he was, in public, " spoke to by the King and Queen."

> " Enter Sir Joseph, gladdening Royal eyes.
> What holds his hand ? a Box of Butterflies !
> Grubs, nests, and eggs of humming-birds to please,
> Newts, tadpoles, brains of beetles, stings of bees. . . ."

There are several entire pieces addressed to Sir Joseph, as, *An Ode on the Report that Sir Joseph Banks was made a Privy Councillor*, very amusing ; and *Sir Joseph Banks and the Emperor of Morocco*. *Peter's Prophecy* is a third short poem (with frontispiece, by Rowlandson), altogether coarse and reckless ; a production that must have tasked heavily the good-humour of Banks's friends.[1]

Thus the Pigmy ; taking measure of a giant.

One of the oddest *supercheries* of the day was a pretended diary of the King, in which he is supposed to be portraying the characters, and arraigning the motives, of some of the courtiers.[2] He suspects that each one has his own axe to grind. For example :

[1] *Peter's Prophecy ; or the President and Poet, or an important Epistle to Sir Joseph Banks on the approaching election of a President of the Royal Society* (London, 1788). The scene portrayed by the artist is a Club dinner. Five guests are feeding on reptiles and other trifles. A servant brings in a young alligator upon a dish. The chairman's knife is ready, and his face anticipatory of impending joys.

[2] *Royal Recollections on a tour to Cheltenham, Gloucester, Worcester, and places adjacent, in the year 1788* (7th ed., London, 1788).

" That Banks is an odd animal ; with the gait and physiognomy of a savage ; with a fortune to render him independent, and with a love of knowledge to occupy him. He is a servile courtier. A spy on the philosophical world, —he enables me unperceived to direct my influence against impertinent and innovating genius ; and I have, by his means, facilitated foreign negotiations, and gratified the vengeance of German despots on men of letters who have taken refuge in England." Further to the effect that Lord Mahon would have been President of the Royal Society but for Pitt's intriguing.

This brochure was evidently inspired by the Prince's party against the Court. The writer was the Rev. David Williams. It was issued by the notorious Captain Thomas Morris, with a short sketch of Williams's life. Morris says that the *Royal Recollections* "have a species of wit and humour, *as vehicles of important truths*, and a facility and happiness of expression which are peculiar to our author." This is equivalent to a flat confession that the slanders and insinuations were done on authority, as part of the campaign against the King and his friends.

A few years later, James Gillray essayed a sportive picture of Banks. It is reproduced on the adjoining page. It will be noticed that there is some grace about it all. The idea that this particular butterfly was specially attracted by the Crown is absurd enough ; but there were many ill-natured people at the time who regarded Banks as something of a toady. The Legend runs thus :

" The Great Sea Caterpillar, transformed into a Bath Butterfly.

" Description of the new Bath Butterfly, taken from the Philosophical Transactions for 1795. This Insect first crawled into notice from among the weeds and mud of the South Sea; and, being afterwards placed in a warm situation by the Royal Society, was changed by the heat of the sun into its present form. It is notic'd

The great South Sea Caterpillar, transform'd into a Bath Butterfly.

Description of the New Bath Butterfly, taken from the Philosophical Transactions for 1795.— This Insect first crawl'd into notice from among the Weeds & Mud on the Banks of the South Sea, & being afterwards placed in a Warm Situation, by the Royal Society, was changed by the heat of the Sun into its present form.— it is notic'd & valued Solely on account of the beautiful Red, which encircles its Body, & the Shining Spot on its Breast; a Distinction which never fails to render Caterpillars valuable.

Sir Joseph Banks
Created a Knight of the Bath by
George 3rd

J. Gillray

A BATH BUTTERFLY

and valued solely on account of the beautiful Red which encircles its body, and the shining spot on its breast ; a distinction which never fails to render caterpillars valuable." Banks had just been made a K.B.

When Sir Joseph had occasion to rebuke any one, he stated his case without heat and without exaggeration. But he did not mince matters. Nor was there any anticipatory excuse or palliation. The really great-natured man does not proceed to extremities until compelled ; and only at that point because any sense of personal injury is subordinate to a sense of outraged principle. It is this latter feeling that creates surprise, and even resentment, in persons who have supposed that they can go to any length they like in imposing on good nature. Banks was frequently being imposed upon, either by friends and acquaintances, or by outside adventurers. There was really, in his case, a premium on presumption.

Mr. Pennant, in his Autobiography,[1] speaks of a rupture which occurred between himself and Banks, but which presently blew over. It was caused by the misconduct of J. F. Miller, the artist of Banks's Icelandic plants, and the occasional draughtsman of flowers for him. Banks had always treated Miller very liberally. But the day came when he must needs dismiss him from friendship.

Sir Joseph Banks to Thomas Pennant.

" Soho Square, *May* 4, 1783.

" Dear Sir,—On my return to town after the Easter holidays I received a message from you by Mr. Dryander informing me that you had purchased from Miller, whom I took with me as my articled draughtsman to Iceland, certain drawings of that country, and enquiring if I had any

---

[1] *Literary Life of the late Thomas Pennant, Esq.* (London, 1793).

objection to your publishing some of them. In answer to
which I must inform you that, without entering into the
question of the propriety of buying things circumstanced
as these were, that I have always considered them as stolen
from me in a most unhandsome and illegal manner ; and
have held myself ready to prosecute Miller if he should
publish them in any shape whatever. So circumstanced,
I trust that you will excuse me for refusing my consent
to their publication, as I must consider such a measure
a material injury to the Journal of my Icelandic voyage
obtained at no small expense ; and probably, if leisure
allows me hereafter to print it, to be attended with no
small emolument. Your faithful servant," etc.

Mr. Pennant did not reply to Sir Joseph in the proper
spirit. There was no pretence of apology for his false
step. He only thought, it would appear, of the success
of his own project. This is much to be regretted, since
the two men had been attached to each other for a good
many years. The present episode is, doubtless, the one
in question, just alluded to.

Before presenting Mr. Pennant's letter, it will be interesting to see the initial cause of the breach between
the two friends.

John Frederick Miller to Mr. Banks.

" COWLEY STREET, *May* 18, 1776.

" SIR,—At the time I was engaged with you to draw
your plants you was so obliging as to lend me the volume
of Redinger's Curious Animals. Herewith I return the
same with my thanks. The bearer of this will also deliver
you the picture which you commissioned me to sell for
you.

" When I was with you last Thursday, the 16th instant,
I was extremely surprized of your treatment to me. Now

that surprize is over, I will answer your accusation with coolness and candour. You, Sir, did reproach me for exhibiting in the artists' room in the Strand the *Massonia.* I gave you then my reasons why, because I did see you withheld the name of those who made the drawings for you. I confess I thought it, and on that account, Sir, I did exhibit that Plant. You, Sir, was also pleased to say you should not be surprized to see half your South Sea plants published by me in time ; and that if I did dare publish, draw, or sell any of them, you would prosecute me to the utmost of the law.

" I am sorry, Sir, you could descend to make that expression. Time, and my actions, will convince you that I am above the law, for I am determined never to transgress it. In the proposals for my intended work you will read that all the plants are to be drawn from the life (not from dry'd specimens). Time will evince, and the public will judge, how far your suspicion is founded in Truth. I am," etc.

The same.

" DORSET COURT, *November,* 1776.

" Mr. Miller's compliments to Mr. Banks. Mr. Miller sends the drawings of the *Heliconia* agreeable to Dr. Fothergill's request. Mr. Miller begs the return of the drawings soon. . . ."

Joseph Banks to J. F. Miller.

" To Mr. Miller, Junr.

" *November* 26, 1776.

" MR. MILLER,—Since you have publish'd engravings of drawings which you made for me at my expense, I desire to have no further concern with you.

" With this you will receive back your engraving of the Chamæleon.

" Mr. Banks will send Dr. Fothergill's drawing of the *Heliconia* back to Mr. Miller when he has done with it."

Mr. Pennant to Sir Joseph Banks.

" DOWNING, *May* 11, 1783.

" DEAR SIR,—A loitering journey prevented me meeting your favour sooner than last night. Be assured I shall never do any act which may deprive you of the least particle of fame or emolument, therefore rest assured that the Drawing returns into the magazine of my private amusement. I was directed to that in question by an information of Mr. Greville that such was on public sale, otherwise should never have thought of them.

" My design is in my preface to my ' Arctic Zoology ' to give an historical, geographical, and physical account of the countries of whose animals I treat ; and my intention was to have illustrated it with views expressive of the regions I passed through. For that end, I applied to Mr. Stephens for the use of a few of the drawings of the headlands on the northern side of N. America, and which he informed me were in your custody. They were those which I saw on your table in company with Captain King. I should have been happy to have had the use of them, as I heard you mention that they were not to be engraven. Nor did I see them in the list of plates shewn to me by Dr. Douglas. Those drawings were mere headlands, very lofty, black streaked with snow. As they are neither with you nor the Admiralty I presume that Captain King has them. My assiduity and labour for so great a length of time make me consider myself as a public man, meriting the assistance of my friends. I therefore hope. Had those drawings been with you I should not have been denied my request. My Preface has had the sanction of good judges, therefore must not be suppressed. I am," etc.

Sir Joseph Banks to Mr. Pennant.

(May 15, 1783.) " SIR,—The description you give me of
Views you are desirous of publishing is so slight, that I am
not able to determine what you mean. Of course I cannot
inform you where they are to be met with. For my own
part, I do not know that any Drawing, the property of
the Admiralty, is lodged in my hands, except such either
have been or are to be engraved in the intended publication of Captain Cook's Voyage. But whatever there may
be, I cannot think it would be proper for me to deliver
them to any one without an order for so doing from the
Board to which they belong.

" Your claim of assistance as a man to whom the
public is under obligations, as far as it relates to the
Admiralty, they must discuss. For my own part, I can
easily settle my opinion upon it. At least, in the present
instance, I am not mistaken in saying it does not entitle
you to expect from me the use of another's property till
you have obtained the consent of the owner.

" You say that your Preface must not be suppressed :
words to me unintelligible, unless they mean a threat.
Indeed, I know not how to interpret the word ' must '
into any other sense. If that should be the case, I can
only say that I consider myself not incapable of defensive
measures if attacked. If otherwise, the whole sentence
must be consigned to oblivion ; as I do not think the
public likely to patronize any symptoms of pugnacity
which I may exhibit, unless I had ample reason given
me for submitting it to their tribunal. I am," etc.

Another occasion upon which Sir Joseph asserted
himself is best described by leaving the correspondence
which follows to speak for itself. Mrs. North, it should be
stated, was wife of the Bishop of Winchester.

Mrs. Brownlow North to Sir Joseph Banks.

"WINCHESTER HOUSE, CHELSEA, *July* 27, 1787.

" Mrs. North presents her compliments to Sir Joseph Banks, and has the honour to send him thirteen of Mr. Brown's drawings, which he subscribed for at Mrs. Greville's when she had the pleasure of meeting him at dinner. There are three more drawings due to Sir Joseph, and it will be unnecessary to pay the remainder of the subscription, which is two-and-a-half guineas, until Sir Joseph hath received the whole. As to the letter-press mentioned in the proposals, Mr. Brown finding he has lost amazingly by the drawings alone, it will not be in his power to comply with that part of his agreement ; but as he is willing to refund the first payment of two-and-a-half to any of his friends who shall disapprove of this deviation from his original plan, Mrs. North flatters herself that Sir Joseph Banks will not be displeased with this alteration."

Sir Joseph Banks to Mrs. North.

" MADAM,—As my inducement to subscribe to Mr. Brown, besides the honour of obeying your commands, was entirely the hopes of promoting the publication of a book which, by being in the hands of different people, might promote the communication of their ideas to each other on the subject of the plants it was to treat upon, I shall certainly accept of his proposal of giving up half his subscription on account of not fulfilling the proposals by which he bound himself ; but as the poor man represents that he is already a loser by the execution of this part of his plan, he intends to adhere to, I am most ready to return the thirteen drawings you did me the honour to send to me yesterday, leaving the 2½ guineas he has already received with him as a homage to your recom-

mendation ; and shall leave orders that they be delivered
to him whenever he will give himself the trouble of calling
for them in Soho Square."

Mrs. North to Sir Joseph Banks.

" FARNHAM CASTLE, *August* 9, 1787.

" SIR,—I had the honour of your letter yesterday, and
am extremely concerned that I should be under the
necessity of giving you the trouble of receiving a second
letter from me. But as there seems to be a misunderstanding relative to the drawings in question, I must beg
to refer you to the paper I gave you of Mr. Brown's
proposals, in which you will see that the letter-press was
only to contain an illustration of light and shade, and no
Botanical description to promote the communication of
his ideas to others. As I had received the two-and-a-half
guineas from you, I thought it right to state the case as
it really was, especially as I had acted the same by the
Duke of Northumberland and others. But I by no means
intended to impress you with sentiments of compassion
for Mr. Brown, but merely to give you the option of
taking the drawings or not as you thought proper; as
he does not stand in need of pecuniary assistance, though
he may want friends to bring him forward in the world
and do justice to his talents. In this light I solicited your
protection, and as it was with no other motive I must beg
to decline your offer of the two-and-a-half guineas. I
shall send the drawings and leave the money due to you
with your servant in Soho Square. I am," etc.

Sir Joseph Banks to Mrs. North.

" MADAM,—The words of your last favour certainly
impressed me with the idea that Mr. Brown was in some
need of pecuniary assistance : which, misled by that
opinion, I offered willingly. I rejoice, however, to hear

188 THE LIFE OF SIR JOSEPH BANKS

that his circumstances put him above the acceptance of my offer. For the matter of his talents as an Artist, I must plainly confess that I do not rate him so high as by your note now before me it appears you do. If any of my friends who may rest their opinion on my judgment in botanical matters had been misled into a belief that I ever conceived an idea that Science would receive benefit from his pencil, which in my estimation is better adapted to trace patterns for the Loom or the Needle than to copy the intricate though apparently simple forms which nature assumes in the vegetable world.[1]

" Excuse, madam, I request, the plainness of this declaration ; which the intention you signify of returning the money I have subscribed makes necessary. Believe me, with due respect," etc.

Dr. Beddoes [2] was a distinguished physician and chemist. He was one of the first to suspect that the recent discoveries in chemistry could be well applied to the study of medicine. He supported his views with much eloquence and lucidity, and had a numerous following.

In 1790 he was reader of Chemistry at Oxford ; but the freedom of his opinions on political topics, especially favourable to the French Revolution, made his position there untenable. He resigned in 1792, being then about thirty-two years of age. Soon after this period his thoughts were turned to what he called Pneumatic Medicine, i.e. the driving out disease by the inhalation of a medicated atmosphere. Proposals were made for raising a hospital at Bristol for systematically carrying

[1] Incomplete (from a draft).
[2] Thomas Beddoes, M.D. (b. 1760). His abilities at Oxford University were early recognized. He was a good linguist, botanist, and chemist, and became known as an enthusiastic philosopher, with vigorous and independent mind. He married a daughter of R. L. Edgeworth, and became the father of Thomas Lovell Beddoes the poet.

out the idea, and much encouragement was offered by Beddoes's friends.

In due course of time, applications came to Sir Joseph Banks. In reply to Georgiana, Duchess of Devonshire, who warmly urged the merits of the proposed Pneumatic Institute, he wrote as follows :

(November 30, 1794.) " Sir Joseph Banks presents his most respectful compliments to the Duchess of Devonshire, and begs Her Grace to give him credit when he assures her that he feels infinite regret in not finding himself able to obey Her Grace's commands respecting Dr. Beddoes. Sir Joseph had once his doubts concerning the propriety of his giving public countenance, of any kind, to a man who has openly avowed opinions utterly inimical to the present arrangement of the order of Society in this country ; but the Duchess's better opinion of the Doctor has wholly satisfied him on that head. His doubts are now confined to the Doctor's project of trying the effect of gases upon patients labouring under the consequences of pulmonary disease. On that head, Sir Joseph is of opinion that there is a greater probability of a waste of human health, if not of life, being the consequence of the experiment ; than an improvement in the art of Medicine being derived from its results. He cannot therefore, with a safe conscience, give encouragement, either public or private, to an undertaking in his opinion more likely to be attended with mischievous than beneficial consequences."

The lady, in the course of her rejoinder, says that her passion for Natural History, Chemistry, etc., had begun abroad, and she was ignorant of the existence of such a person as Dr. Beddoes until she was recently introduced to him at Bristol. " It was not till some time after that I heard his political opinions had occasioned his leaving Oxford. But I was told, at the same time, that he had

long since abjur'd them and was wholly given up to
philosophical experiments. . . . My idea was that his
plan was a fair and open trial at real investigation ;
and that by proving once for all whether pneumatic
Chemistry was, or was not applicable to medicine, he
would either open up a new branch to general utility,
or for ever prevent any trial that might be hurtful. . . .
I think your protection of such consequence to any undertaking that I shall feel inclined to advise Dr. B. to give
up any public trial without it."

Sir Joseph Banks to the Duchess of Devonshire.

"December 2, 1794.—Sir Joseph Banks presents his
respectful compliments to the Duchess of Devonshire.
He once more entreats Her Grace to believe him, when
he assures her that he has felt more uneasiness than he
can express from the misfortune of differing in opinion
from Her Grace relative to the probable event of Dr.
Beddoes's experiments ; and would have instantly sacri-
ficed his reason to his respect for Her Grace, had he not
found it necessary to be very careful in preventing his
name from appearing as an encourager of projects he
does not think likely to prove beneficial to Society. . . .

" . . . he will readily change his opinion the moment
he is enabled to do so by a conviction of his having before
been in the wrong. He cannot suppose his name a matter
of material consequence to Dr. Beddoes, being in no
degree connected with the study of medicine. . . . He
feels confident that the signature of Sir L[ucas] P[epys]
or any other of the fashionable physicians would do the
Doctor a hundred times more service than a dissertation
from him, in favour of pneumatic medicine.

" Sir Joseph would not have been a moment after he
heard of the Duchess's arrival in London, without presenting himself at Devonshire House, had it not been

the week of the anniversary of the Royal Society, when he is necessarily so much occupied. . . . He will certainly to-morrow pay his respects where they are so justly due."

A few days afterward, there was a similar interchange of communication with Mr. James Watt, between whom and Sir Joseph there was much friendly intercourse. But Banks was not to be moved. He had seen no proofs of the efficacy of the new system ; and he ventured to repeat the suggestion that " the medical gentlemen, many of whom are in the most affluent circumstances, should lead the way " by their subscriptions and by their moral support. . . . " You flatter me by saying that my name will be of use to Dr. Beddoes. If it is capable of being useful in attracting other subscribers it is because I have been careful hitherto not to annex it, in any matter of importance to humanity, unless I had sufficient conviction of the probability of its success. In the case of Dr. Beddoes I do not fully understand it. . . . I hope, Sir, this argument will satisfy you that I am not unreasonable in requesting that I may not be pressed any more by the Doctor's friends to do what I have already formally declined to do."

By the help of James Watt, R. L. Edgeworth, and other liberal friends, the Pneumatic Hospital was opened at Bristol in 1798, and lasted a few years.

CHAPTER XII

EUROPEAN FAME

SIR JOSEPH BANKS was not only mindful of the courtesies due to foreign *savants* from their friends in England. He extended toward them a hand full of zeal and warmth. The prevalent knowledge of this fact had much to do with the respect in which he was held throughout Europe, and the reverence usually paid to the Royal Society.

One incident that occurred during the period of the French Revolution is so characteristic of him that it may be presented with some detail.

The explorer, La Pérouse was long overdue. He was last seen at Botany Bay early in 1788. The National Assembly at length sent out a search Expedition, under the command of Joseph D'Entrecasteaux, in September, 1791. He was accompanied by the botanist La Billardière, charged with the natural history department of the Expedition. This latter gentleman was well known to Banks, in whose house he had spent much time studying the collection of plants, minerals, etc., stored in Soho Square. At the request of La Billardière, Sir Joseph wrote a series of instructions for his use, politely disclaiming, meanwhile, that it was in his power to teach him anything new.

The voyage of D'Entrecasteaux was in every way successful, except as to finding any trace of La Pérouse. He had reached the neighbourhood of Java with hopes of a speedy end of his long cruise. But here misfortunes

began. The Captain succumbed to a complication of
scurvy and dysentery. The command fell upon De
Rossel, who had the ill-luck to discover that it was war-
time. His ships and his treasures were seized by a
British frigate, and La Billardière's valuable collection
of plants, birds, etc., was brought to England. The
Duc d'Harcourt, an *émigré* officer in the British service,
appears to have been the captor ; and on his arrival in
London he proposed to the Queen's Chamberlain that the
fine Herbarium now in his hands should be presented
to Her Majesty. Major Price forthwith wrote to Sir
Joseph Banks, asking him to visit Harcourt House and
inspect the collection with a view to determine whether
the whole, or a part, should be added to that of Her
Majesty.

Sir Joseph Banks to Major William Price.

" SOHO SQUARE, *March* 31, 1796.

" MY DEAR SIR,—In consequence of my sending on
Tuesday morning to Harcourt House, in order to learn
whether orders were left for my having admission to see
the collections there during Lord Harcourt's absence, I
received a letter from the Duke . . of the whole
curiosities collected on board the Discovery ships that
were commanded by M. D'Entrecasteaux, and that they
are offered as a present to the Queen by the King of
France.[1] . . . They consist of a vast Herbarium, collected
in all the places at which the ships touched, a large
collection of dried Birds, a considerable number of dried
Lizards and Snakes, some Fish in spirits, and some Insects
which are said to be much damaged.

The collection of plants bears testimony to an industry
all but indefatigable in the botanists who were employed,

---

[1] i.e. Louis XVII, who was, by a legal fiction, the sovereign of the
d Harcourts.

o

the chief of whom I am sorry to say was the principal
fomenter of the mutiny which took place in the ships,
built upon the strongest Jacobin principles. . . .[1]

" As the duplicates would serve no purpose but to
encumber Her Majesty, I shall with pleasure, if I am
honoured with Her Royal commands, undertake to
select a complete collection of one good specimen of
such species ; but as individual specimens from whence
they are to be taken cannot number less than ten thousand, every one of which must be separately examined,
I dare not undertake to complete the work in less than a
year from the present time.

" The Birds are in tolerable preservation, and many of
them extremely beautiful. If Her Majesty chooses to
make a collection, I would by all means advise Her
Majesty to accept them. There are about twelve hundred,
many of them quite new. The cost, however, of stuffing
them . . . and providing glass cases to contain them,
would amount to several hundred pounds. The Snakes
and Lizards are dried. They might be mixed with the
Birds, as some of them are not only curious but as handsome as such reptiles can be. . . .

" Provided, as I conclude is likely to be the case, that
Her Majesty does not choose to encumber herself with
the stuffed animals, a word from her would probably
direct the Duc d'Harcourt's attention toward the British
Museum, where they would become a National ornament,
and promote materially the knowledge of Natural
History. . . ."

Further, there is a suggestion that the charts and
sketches would be best placed in the King's library,
" where men of real Science have always access, in a

[1] It was a curious complication of affairs. The naval officers were
adherents of the White Flag, and the naturalists were all republicans.

manner that does honour to the King's liberal mind
and the proper discrimination of his Librarian."

The spoil thus disposed of, Sir Joseph had probably
begun his self-imposed task of examination and selection, graciously accepted by the Queen. Five months
elapsed. Then representations were made, apparently
from Jussieu, which caused Banks to look at the recent
windfall in a new light. The French Directory wished to
rescue the Prize on behalf of their nation. There were
those in England who objected to the restoration of the
Collection, and used the strong language of the period,
about Regicides and Cockatrices; but they mostly
yielded to Banks's opinion on the subject. The following
excellent letter tells the whole story, from his point of
view.

Sir Joseph Banks to Major Price.

" *August* 4, 1796.

" MY DEAR SIR,—Since I had the honour of writing to
you on the subject of the Collection of Curiosities offered
by the Duc d'Harcourt as a present to the Queen, the
whole of the business relating to these things has taken a
very different form. I sincerely hope it will not be productive of any disappointment to Her Majesty, and I
feel it my duty to do all in my power to obviate as much
as I am able all possibility of that being the case.

" When the Collection was offered to the Queen it was
supposed by all who were concerned in making the offer
that it belonged to the present King of France ; and it
was believed that the late King interested himself
personally in directing the outfit of the voyage, and that
His Majesty actually employed on his own service the
persons engaged to make Collections. An application
having since been made by the Directory of France,
requesting that the collection might be returned to
M. de Billardière, in the same manner as M. Ulloa's papers

were returned to him when captured by an English vessel, in order that he might be enabled to publish his observations for the advancement of knowledge as Ulloa had done. For, the enquiry was made, in the course of which it appeared that the late King of France took particular interest in the outfit of this Expedition in which M. de La Billardiere sailed; and that the mistake of His Majesty's having done so probably originated in the fact of his having done so in the voyage of the unfortunate M. de La Pérouse : who, it is confidently said, was honoured with private instructions in the King's own handwriting.

" Under this view of the business, His Majesty's Ministers have thought it necessary, for the honour of the British Nation and for the advancement of Science, that the right of captors to the Collection should on this occasion be waived ; and that the whole should be returned to M. de La Billardière, in order that he may be enabled to publish his observations on Natural History in a complete manner. . . . By this Her Majesty will lose an acquisition to her Herbarium which I very much wish'd to see deposited there ; but the national character of Great Britain will certainly gain much credit for holding a conduct toward Science and scientific men liberal in the highest degree.

" Have the goodness, my dear Sir, to state these matters to the Queen on the first convenient opportunity, and do me the further favour of acquainting me as soon as you can with Her Majesty's pleasure on the subject. Add, if you please, that it is in my power, I verily believe, to make an addition to Her Majesty's Collection as valuable at least as the one in question. That I shall feel myself honoured in the extreme if I have permission to do so ; and that in case I am so fortunate I will, the moment I return from my annual journey to Lincolnshire, undertake the business of preparing and arranging the plants in such manner as I think the most likely to render them worthy the honour I solicit for them and for me."

Sir Joseph further wrote to Jussieu, telling him that the British Government had, after much pressure, consented to waive belligerent rights in the present case. The following remarkable passage is, we conceive, biographical matter : a confession worthy of being placed along with other testimony to Banks's frank and generous character :

" I confess I wished much to have from his specimens some of those discoveries in the natural order of plants which he must have made ; but it seemed to my feelings dishonourable to avail myself even of the opportunity I had of examining them. . . . I shall not retain a leaf, a flower, or a botanical idea of his. I have not possessed myself of anything at all of his that fortune committed to my custody."

Banks's conduct on this and similar occasions was never forgotten, when an orator sought inspiration for a eulogy of British generosity. After the death of La Billardière [1834], the story of this adventure was once more told.

The affair of the naturalist Dolomieu gave another occasion for Banks's intervention, which he was enabled to offer through his old friend Hamilton.

Sir William Hamilton was an attached friend of Banks, for perhaps thirty-five years. When in residence at the Court of Naples, he was a frequent correspondent. His letters range over such matters as volcanoes and natural curiosities, vases, books, politics, gardening, and Emma Hamilton ; together with minor topics, as the defective points in the character of the Neapolitan Court and people, and repeated invitations of Banks to visit Italy. He was on leave in England in the autumn of

1783, and took an active part in the proceedings of the Royal Society : doubtless he was one of the stalwarts at the service of the President during that winter. Hamilton was proud of his membership of the Royal Society, and he certainly contributed something toward its glories. We owe to him much practical knowledge of volcanoes, furnished at a period when the topic was matter for elementary speculation.

Soon after Sir William had returned to his post in South Italy, he was seized with the idea of forming an English garden, adjacent to the country palace of Caserta ; and suggested to the Queen the great improvement it would make to the estate. Banks was already a familiar name to the Royal couple, and several petty compliments were exchanged. Sir William having pointed out the advantages of Banks's advice and assistance, the Queen was readily induced to consent to the project. She promised to allot £100 sterling per month toward the expense, and urged the choice of a British gardener and nurseryman.

Sir William writes to Banks : " I told her that you had been so good as to promise to assist me. . . . A man of sense and judgment, and high in his profession, must certainly be tempted, to quit his country and establish himself in a distant foreign one. He should therefore be liberally paid. . . . All this is to be done without the King's interference ; and she rejoices in the thought of surprising him some day with a plate of fruit out of her garden much superior to his. And, indeed, that may very easily be done ; for, notwithstanding the great advantage of climate, we have in England (except grapes and figs) every fruit of a superior quality to that of Naples. . . ."

" My collection of visiting-tickets for Miss Banks increases daily ; and indeed some are very curious. I will send what I have by the first safe hand. They are too precious to be trusted by sea."

This last sentence concerns an undertaking which Banks appears to have entrusted to his friends all over Europe. Allusions will be made later on to this monster Collection.

Further reference to the projected garden shows that Hamilton was taking great pleasure in the new occupation it gave him. " The whole art of going through life tolerably, in my opinion, is to keep one's self eager about anything," he says. He really did keep eager about the Caserta garden. It caused him infinitely more trouble than he could have expected ; but it resulted in a great triumph of skill and taste. The gardener whose services were obtained by Banks was John Graefer, a German who had come into England when very young and was now quite naturalized. He was with Philip Miller and other good horticulturists; and afterward raised a nursery-garden at Mile End in partnership with two other adepts.

Graefer had enormous difficulties to meet, such as racial opposition and individual national laziness. The obstacles Sir William had to contend with were the jealousy of courtiers, and of the household servants who beheld some of their perquisites diverted from their natural course. After a long lapse of time, King Ferdinand became interested in the garden, took it into his own hands, and showed by his treatment of Graefer that the man was to be properly honoured. Then things went very well. Hamilton presently writes with a good deal of enthusiasm. The growth of all the imported plants (mostly from England) was surprising. Exotics were everywhere flourishing. " The Camphor-tree has shot up, in the open air, six feet in three months. The verdure of our turf at this moment (August) is as fine as in spring, seeing we have command of water."

The second marriage of Sir William had been undertaken after something like a consultation with his friend

Banks. Whether there was anything of the nature of reproach on his part, we can but guess. The following letter seems to show Sir William somewhat in an attitude of defence. The unconsecrated union with Emma Lyon had now existed some time. Banks, who belonged to a class of persons whose propriety was unimpeachable, but who were not willing to quarrel with others whose notions were less strict, as long as the outward decencies were observed, had evidently suggested marriage. Sir William was greatly attached to him, and bore with good humour this friendly interference. The question being raised, however, Banks had the pleasure of welcoming his friend in London in the following year. The marriage took place in September, 1791, at Marylebone church.

" NAPLES, *April* 6, 1790.

". . . To answer your question fairly : was I in a private station, I should have no objection that Emma should share with me *le petit bout de vie qui me reste*, under the solemn covenant you allude to ; as her behaviour in my house has been such for four years as to gain her universal esteem and approbation. But, as I have no thought of relinquishing my employment, and whilst I am in a public character, I do not look upon myself at liberty to act as I please ; and such a step I think would be imprudent and might be attended with disagreeable circumstances. Besides, as amidst other branches of Natural History I have not neglected the animal woman, I have found them subject to great changes according to circumstances, and I do not like to try experiments at my time of life. In the way we live we give no scandal ; she with her mother, and I in my apartment, and we have a good society. What, then, is to be gained on my side ? It is very natural for her to wish it, and to try to make the people believe the business done, which I suppose has

caused the report in England. I assure you that I
approve of her so much that if I had been the person
that had made her first go astray, I would glory in
giving a public reparation, and I would do it openly.
For, indeed, she has infinite merit, and no princess could
do the honours of her palace with more ease and dignity
than she does those of my house. In short, she is worthy
of anything ; and I have and will take care of her. But
as to the solemn league, *amplius considerandum est.*"

" Now, my dear Sir, I have more fairly delivered you
my confession than is usually done in this country, of
which you may make any discreet use you please. Those
who ask of mere curiosity I should wish to remain in the
dark. Adieu, and believe me ever sensible of your friendship and kindness to me."

The memorable step taken by Sir William, in 1791,
was abundantly justified if personal comfort and mutual
happiness are worth anything. His letters to Banks,
gossiping over all topics political and social, are never
wanting in some fond reference to his second wife.
Her mental improvement and her facile acquirement of
the ways of high life are his delight. " She has got the
Italian language better than I have in twenty-eight years."
English ladies of the first rank receive her. . . . " She
is often with the Queen, who really loves her." . . . There
is now no Past. And they have a common gratification
in the friendship of Sir Joseph Banks.

Emma, Lady Hamilton, to Sir Joseph Banks.

" GIARDINO INGLESE, *May* 31, 1797.

" DEAR SIR,—I cannot let Sir William's letter go
without assuring you how happy I am to find His Majesty
does justice to your merit, and that you are one of his

Privy Council. I assure you, Sir William loves you dearly, and has felt happy on hearing of any addition to your honour and comfort. Think, then, as I am inclined to love all his friends, how rejoiced I am at your happiness and welfare ; for indeed I do not consider you as one of his common friends, and therefore love and esteem you more than you think I do. I have not shewn to you much that I have ever thought on you, for I have never wrote you a line, God knows how long ; but you do not imagine how my time has been taken up, and what I have had, and have now, on my hands. Nor indeed, I never could have thought, nor do I feel myself worthy of the happiness and honours I have enjoyed for the last few years.

" We are now at the Queen's garden, resting, and getting Sir William's health against the return of the Royal Family from [? Foggia], for we shall have hot work at the marriage.[1] The Queen has been with us two days, and has made me very happy by giving me a sum for Graefer to go on with the garden. So we are all very busy to get on before the princess arrives, as I hear she is fond of Botany. She must have the garden, she must introduce Agriculture, she must shew the Neapolitans what Graefer is ; and your name with his must go down to the Neapolitan posterity. For, by sending him here, you have given them Taste. God knows, at present they have none ; not one except my dear adorable Queen. You should know her. She is worthy to be known to you and a few others of your character. Do you know Sir Gilbert Elliot ? Ask him what she is : everything that's great and sublime. I hope one day to present you to her. I have told her so, for she admires you from hearsay ; and some day, when I have time, you shall know two or three traits of her excellent character. And, do not

[1] Archduchess Clementina to the Hereditary Prince of the Two Sicilies.

believe the book the infamous Gorani has written against
her : *'tis all false.*

" We go to town to-morrow to prepare for Sunday,
the King's birthday. We have a dinner of a hundred,
and an assembly of a thousand in the evening. In
England, where you count nobility by the hundreds,
we count them by thousands ; for at my ball of a
thousand there will be at least eight hundred Duchesses,
Princesses, etc. ; and not one of them worth a good
farmer's wife in England. However, I wish you was
there ; you would be delighted with the country and
characters.

" I wish I could send you a curiosity I have got,
that you might examine her before the Royal Society.
'Tis an old woman of 112, but she looks so young you
would not give her more than 50. She has neither teeth
nor gums ; she was blind, lame, and wrinkled till she
entered her hundredth year. She then, like a serpent,
changed her skin. Her eyesight came back ; she walked ;
and her skin became as fine as any young girl's ; her
neck, arms, etc., are very round, fine, and hard ; her
voice is like a girl's of ten years old ; she sings like a girl
whose voice is still changing from the girl to the woman ;
her memory is perfect, as she tells me things of the last
century ; not the least deaf ; has been fifty years a
widow ; had thirteen children, all dead of old age ; she
walks ten or twelve miles a day ; if sick, does not eat for
two days, and gets well. We asked her how she had contrived to live so long. Her answer was, ' Signora, non ho
preso mai ne medicamenti ne collera.' Her name is
Mary, but she is called Mariuccia, which is the abbreviation
of Mary, what we should say Polly. She has had a third
set of teeth ten years ago ; but drinking lemonade sour,
they all dissolved, as they were like the teeth of a fish.
She has five [?] a day to live on from the King, and I gave
her to eat. I told her I would speak to the Queen to get

her something more. She said she was content with what
she had. She says she is twelve years old, and when she
is fifteen she will be married. She never counts the last
century, and I think may live to the age of Old Parr.
Her neck is just like a young girl's ; and when I made her
shew it to Sir William, she blushed, and felt so ashamed,
more than any modern young lady would do. Sir
William is in love with her. I have had many physicians
to feel her pulse ; and one, a very learned young man
who is with Prince Augustus, says her pulse is the pulse
of a young girl. She dances very often the Tarantella
with me. When she changed her skin, the people here
got it, and kept it like the bone of a saint, for she is
considered as a miracle. So much for Mariuccia, who
has made me bore you by reading this scrawl, for I wrote
in a great hurry. But am ever, dear Sir, your obliged
and sincere   " EMMA HAMILTON."

The correspondence with Naples had become inter-
mittent, probably because of the disturbed state of
communications between England and the Continent.
It revived with some vigour in the year 1799. A long
letter from Sir William (September 13) is quite in his old
vein ; yet with hints that he is getting worn out, and
tired of his post. There is much self-satisfaction over the
services he has been able to perform for the King and
Queen of Naples, with the assistance of Lord Nelson.
A prodigious number of presents and rewards are in the
air. Every Captain of the squadron has shared in these
gifts. Nelson is now Duke of Bronté with an estate
worth 18,000 ducats per annum. Graefer has been made
Nelson's agent, with a handsome salary and an annuity
for his wife. . . ." We have deferred our departure till
next spring, when alive or dead I shall come home ; for,
at my first wife's particular desire I am to lie by her in
Slebech church when I am dead. We shall roll soon

together into Milford Haven, for the sea is undermining that church very fast. . . ."

In the following November it was Banks's turn for long letters. He had been receiving many applications from foreign *savants*, begging his intercession on behalf of D. Gratet de Dolomieu, then in prison at Naples. Dolomieu, on his way homeward from Bonaparte's Egyptian campaign, was taken prisoner in the dominions of the King of Naples because of some former compromising circumstances with the Order of Malta. It was said he was being treated with barbarity. The scientific world of Europe made urgent demands for release. Hence thirty applications, at least, to Sir Joseph Banks.

Dolomieu was a harmless naturalist, a pioneer in geology, and a famous mineralogist. The thing appealed strongly to Banks, and he pressed the matter upon Sir William's attention, at some length. He wrote also to Lady Hamilton, probably judging that her intimacy with Queen Caroline, of which he had heard so often, would not fail to bear upon the fate of the prisoner.

Sir Joseph Banks to Emma, Lady Hamilton.

" *November* 8, 1799.

" MY DEAR LADY HAMILTON,—I have ever been convinced of your friendly disposition toward me ; but the readiness of my friend Sir William to anticipate on all occasions my wishes has been so constant hitherto that I have never before been under the necessity of putting your ladyship's friendship to the trial.

" Now, however, a business has occurred which cannot, I am convinced, be effected unless both your influences are united in soliciting it. If it is possible to engage the gallant Admiral to join the Trio, which is in the power of no one but yourself, what may I not hope in everything from the warmth of your friendship ?

" What I wish for is the liberation of Dolomieu, who, for the honour of the Kingdom of Naples, ought to be liberated ; because he is, and must be considered as, a prisoner of war, and cannot be made answerable by the law of nations for actions done by him under the orders of his Commander-in-Chief. I need not trouble you with details on the subject, as Sir William will no doubt shew you my letter to him, by which you will find how lively an interest the scientific persons in Europe take in his fate.

" Allow me, my dear Lady, to be indebted to you for a favour which none but yourself can offer. Your influence is, as it ought to be, unbounded. Gloriously as you are now circumstanced, having at least a third part of the merit of replacing the crown on the heads of their Sicilian Majesties, undertake the business with your usual spirit. It will be for the honour of the Court of Naples, for the advancement of Science, for the benefit of humanity, and will fix an indelible obligation on one who has the honour of signing himself," etc.

By stipulation in the treaty of peace, after Marengo, Monsieur Dolomieu was set free. On July 16, 1801, Sir Joseph was able to write to him congratulating him on safe arrival in his own land, and hoping he was able to resume his scientific work. But he died in the following November.

Sir Joseph Banks to the Right Honble. William Pitt.

" SOHO SQUARE, *March* 17, 1797.

" Sir Joseph Banks presents his respectful compliments to Mr. Pitt, and requests the honour of a few minutes' conversation with him at any time that may best suit with Mr. Pitt's convenience. It is on the subject of opening a communication with Paris for the reception of the

Literary productions of the Members of the *Institut National*, and other scientific persons, some of which are highly interesting to the Royal Society ; and for sending in return the ' Philosophical Transactions,' the ' Greenwich Observations,' etc. This has been omitted for some years past. Sir Joseph has an opportunity of doing this now by the means of Mr. Charretié, Commissary here for the exchange of prisoners. . . ."

Sir Joseph Banks to M. Charretié.

"SOHO SQUARE, *March* 18, 1797.

" SIR,—I lose not a moment in returning to you my best thanks for the zealous and effectual steps you have taken to open an intercourse between the *Institut National* and the Royal Society. Such communication cannot but be of material use to the progress of Science ; and may also lay the foundation of a better understanding between the two countries in future than, unfortunately for both, has of late years taken place. . . ."

These two notes illustrate one phase of the disagreeable position in which scientific, and, indeed, many private persons, were placed during time of war. It was the happy lot of Banks to have so much personal influence in Europe, that he was enabled to alleviate many of the undeserved troubles of the time.[1] He was always ready to exert himself in restoring men of science to liberty, as in the case of Dolomieu ; and in recovering valuable collections from the grasp of British cruisers. According to Brougham, no less than ten restorations of property to the *Jardin des Plantes* were made at his instance, " which had fallen a prey to our naval superiority." As time went on, and war was renewed after the Peace of

[1] Lord Auckland spoke of him as " His Majesty's Ministre des Affaires Philosophiques."

Amiens, with the result that many unfortunate civilians were detained in France, it was quite a big department of Sir Joseph's affairs to correspond with persons begging his intervention.

The time came when the Ministry of the day felt it inconvenient to meet these demands on the part of Sir Joseph. Lord and Lady Shaftesbury were long detained in France against their will. They wrote to Earl Radnor complaining that all other methods of obtaining their release had failed, and begged him to procure a letter from Sir Joseph Banks to the French *Institut* requesting their intercession. At this point Banks had become discouraged by learning that Lord Howick seemed to dislike his continued interference. A memorandum, in Banks's handwriting, dated October 4, 1806, runs thus : " Was I countenanced as I think I ought to have been in extending as much as possible the influence I have over the literary men in that country, I think I could have done much. But in that case I must have been allowed some influence in this country, enough to have enabled me to solicit with effect such favours for Frenchmen confined here. This however is now withdrawn by Lord Howick's means, and I am deprived of its power." A year and a half later he mentions that his correspondence with France was stopped by the late Administration, and that the present one refused to allow him to resume it : " I have released ten prisoners, all of whom would but for me have been still in France. The facilities I met with were so much on the increase when Lord Grey [*then* Howick] stopped me, that I could have done a good deal more before this time."

The spoils of war, in the shape of " curiosities " of any sort, sometimes found a home in this country without being reclaimed. This was the case with Joseph Martin's collection, from Cayenne, captured by a Hastings privateer. Seven thousand birds, animals, and plants,

many of the last in a " high state of health and vegetation." The captors proposed to give the King a first
refusal, and afterward offered them to Sir Joseph Banks.
Aiton went down the river to inspect the plants, and
reported favourably; and presently took charge of
them at Kew. In the end a great many of them came
into the hands of the British Museum.

The following unpublished letter from Lord Nelson
tells its own tale. We have not been able to identify the
prize nor to discover what became of her precious cargo.

" *Amphion*, OFF TOULON,

" *July* 9, 1803.

" MY DEAR SIR JOSEPH,—One of our frigates has taken
a French corvette, which has been at Athens and brought
from thence some cases of I know not what; but I
suppose things as choice as Lord Elgin's. I have directed
that the cases should not be opened, but sent to England
as they are; that if the public wish to purchase them,
Government should have the option. As they are the
property of the lowest seamen, they cannot be presented,
or I should have been truly happy to have had it in my
power to give them to the Royal Society or to the
Academy of Arts. However, I have directed them to
you, hoping that you will take the trouble that Government may know the value of them, and if they choose
to buy them pay the money to the Agent. If they do not
think them worthy of their attention, I must content
myself with my kind thoughts toward my country, and
direct the Agent to sell them. Although I do not think
that these works (I suppose of Architecture) are in your
line of pursuits; yet, by addressing myself to you I
have given myself a great pleasure in enquiring after
your health, and of assuring you of my sincere respect
and esteem. . . ."

P

It is beyond dispute that the bitter relations existing between England and France at this time were softened by the attitude of literary and scientific men on both sides of the Channel. The French naturalists, as a rule, were ardent republicans, but they were Naturalists first. The British scientists eschewed politics altogether, content with their traditions and with a national stability which recent events in Europe seemed powerless to disturb. Hence, such a body as the *Institut National* occupied a unique position; between the world and the devil, it may be said. In one way, it was enabled to profit by the prevailing disorders and dissensions, seeing that it afforded an intellectual refuge for those who could not entirely keep away from the storm.

In November, 1801, Count Rumford sent word from Paris to Sir Joseph, to the effect that the *Institut* had elected a number of foreign members, including among them Banks, Cavendish, Herschel, Maskelyne, and Priestley.

Sir Joseph Banks to the President and Secretaries of the National Institut of France.

" LONDON, *January* 21, 1802.

" CITIZENS,—Be pleased to offer to the *Institut* my warmest thanks for the honour they have done me in conferring upon me the title of Associate of this learned and distinguished Body.

" Assure at the same time my respectable brothers that I consider this mark of their esteem as the highest and most enviable literary distinction which I could possibly attain. To be the first elected to be an Associate of the first Literary Society in the world surpassed my most ambitious hope, and I cannot be too grateful toward a Society which has conferred upon me this honour, and toward a nation of which it is the literary representative.

A Nation which during the most frightful convulsions of the late most terrible revolution never ceased to possess my esteem ; being always persuaded, even during the most disastrous periods, that it contained many good citizens who would infallibly get the upper hand, and who would re-establish in the heart of their country-men the empire of virtue, of justice, and of honour.

" Receive more especially, citizens, my warmest acknowledgements for the truly polite manner in which you communicated this agreeable intelligence."

This incident caused an unexpected uproar, as soon as Banks's letter appeared in the London newspapers. He and his friends were amazed to find that the occasion was to be seized for disturbing his hold upon the Chair of the Royal Society. According to Brougham, Bishop Horsley took part in the attack. The most widely-read effusion on the topic was probably that inserted in Cobbett's *Political Register*.[1] The tomahawk lately imported from America was already popular, and the *Register*, as its medium, was in its vigorous youth. Whenever there was a grievance, real or imaginary, Cobbett's energies were certain to be seconded by some one as daring and little scrupulous as himself. A Gallo-phobe correspondent (signing himself Misogallus) seized the opportunity for a fierce attack on Sir Joseph, re-producing his letter in full, with a torrent of abusive sentences in condemnation alike of its purport and of its manner : " Load of filthy adulation " ; " So little honour-able to your character and so insulting to the Society over which you have long presided, that an explanation or disavowal is demanded by the public voice " ; " Senti-ments a compound of servility, disloyalty, and falsehood." And the date, January 21, anniversary of the murder of Louis XVI, was unfortunate ; it must have been designed

[1] Vol. I (1802).

" with an opportunity of wounding the pride of Englishmen," etc. Together with some broad misstatements of fact. As a matter of taste, Banks's letter is by no means beyond criticism, especially in its allusion to " disastrous periods." There was certainly cause for the enemy to blaspheme.

However, nobody was one penny the worse. Sir Joseph's cronies laughed both at, and with, him. His old friend J. Lloyd reports that no less than seven copies of the *Register* number had been sent to him in his Welsh retreat, from different friends.

As the opening of the winter session of the Royal Society approached this affair was revived with some energy. " A Fellow of the Royal Society " wrote to the *Political Register* (November 4) with a torrent of angry abuse and sneers about the recent insult to the Society, with this conclusive declaration : " If to be one of the forty foreign members of the *Institut* be better than President of the Royal Society, let us find some other person on whom to confer a favour long considered of the highest order." So the murder was out. The small anti-Banks faction wanted a new President. Sir Joseph was re-elected in due course.

Misogallus (who was never identified) fired a parting shot in the *Register*, December 7, to which he subjoined mock congratulations to Banks on his being once more seated in the Presidential Chair. According to a letter of Sir J. E. Smith, the attack in Cobbett's *Register* was the handiwork of either Mr. Windham or his friend E. A. Woodford, the Paymaster for *Émigres'* allowances, who had been offended by Banks's drastic conduct over the La Billardière collection of plants. Others supposed it to have come from Bishop Horsley's pen.

CHAPTER XIII

THE FOUNDING OF AUSTRALIA

A NEW world was now coming to its birth, far away in the Southern Seas.

It has not hitherto been certainly known to whom is to be credited the proposal to colonize New South Wales.[1] From the following considerations, there can be no doubt that it originated with Sir Joseph Banks, either in his own mind, or in conference with two or three friends. Botany Bay was associated with a glorious period in his life; when, in company with Solander and Cook, and obsessed with a furious zeal for his favourite science, he visited a new and fertile and health-giving land. He could never forget Botany Bay. Its treasures were represented in his Herbarium, and many of his comrades on the Expedition were still among his personal friends.

A Committee of the House of Commons was sitting in 1779; occupied on the question of the transportation of felons. Sir Joseph Banks was examined before this Committee. He urged that the coast of New Holland was the place best adapted for the purpose; the country was thinly peopled, the climate was mild and moderate,

[1] The people of Sydney have been sufficiently enterprising of late years in telling the tale of their early history. They have been the better able to do this, by the acquisition of a number of documents of all kinds which were bought at Sotheby's sale of Banks MSS. in 1886. Several volumes of " Records " have been printed ; and the beginning of an authoritative History of New South Wales is the work of Mr. G. B. Barton, Barrister-at-Law (vol. I, 1889). See also J. H. Maiden's *Banks the Father of Australia* (Sydney, 1909).

and the soil had sufficient variety for purposes of cultivation ; wild beasts were few, fish was plentiful, grass was
luxuriant, and the water-supply good. He proceeded to
enlarge upon the means necessary for assisting an infant
colony ; and upon the ultimate advantages which would
fall to Great Britain, in new outlets for mercantile
activity and maritime enterprise. The Committee in their
Report asserted the humanity and soundness of Sir
Joseph's views.[1]

When the conflict with our American colonies finally
ended in separation, in 1782, a vast number of the people
had already been compelled to find refuge in other lands.
Many of them went across the frontier, and became the
staple foundation of modern Canada. Many more came
to the old home. England was swarming with these
Loyalists, as they were called. They were often aided
from the public funds ; but the question arose, and
became urgent, as to their ultimate settlement. And
there arose a project, for those who were willing to go,
for making a new home in the land discovered by Captain
Cook and immortalized by Banks and Solander.

One of Banks's old shipmates was J. M. Matra (otherwise Marra), already mentioned in connection with the
second cruise of the *Resolution*. He was probably an
Irish-American of those days. In 1783, he wrote to
Lord North, asking for a " share in the allowance
granted to the Loyal Americans."

In July of the same year, he sent this remarkable
message to Sir Joseph Banks :

" I have heard a rumour of two plans for a settlement

---

[1] The proposal to make a convict settlement in New Holland was
not everywhere approved. Visions of an Alsatia, a nest of pirates and
so forth, arose in the eyes of many persons. Alexander Dalrymple made
a strong protest (July, 1785) to the Hon. East India Company against
establishing a penal colony in the South Seas.—*v. A Serious Admonition
to the Public on the intended Thief Colony at Botany Bay*, being comments
on Dalrymple's letter ; the same being quoted in full.

in the South Seas ; one of them for New South Wales to be immediately under your direction, and in which Sir George Young, Lord Sandwich, Lord Mulgrave, Mr. Colman, and several others are concerned. . . . I have frequently revolved such plans in my mind, and would prefer embarking in such to anything that I am likely to get in this Hemisphere." [1]

Mr. Matra seems to have become the voice of this casual committee of Sir Joseph Banks, Sir George Yonge, and others. He published (under date of August 23) *A Proposal for Establishing a Settlement in New South Wales*, the gist of which is as follows :

The project would, in course of time, atone for the loss of our American colonies. The climate was favourable, and the country offered inducements for adventure and exploration. " It would afford an asylum to those unfortunate American loyalists to whom Great Britain is bound by every tie of honour and gratitude to protect and support, where they may repair their broken fortunes and again enjoy their former domestic felicity," etc. Details of plans for carrying out the project were suggested for consideration. The scheme was approved by many intelligent Americans. Lastly, the writer eulogized the liberal and splendid spirit of Sir Joseph Banks, who was prepared to support his opinion in favour of it all with His Majesty's Government when they chose to consult him. [2]

That the idea was popular is manifest by Matra's concluding remark, that he was pestered with letters about it. The Coalition Government was approached, but Fox and North went out of office in December, and the thing was shelved. Early in 1784 the new Ministry gave it some attention. Lord Sydney proposed to improve the plan by connecting with it a convict settlement,

[1] Addl. MSS., 33977/206
[2] *Historical Records of N.S. Wales*, vol. I, pt. 2, xxv.

in lieu of Western Africa, whither felons had been transported during recent years. The opportunity of sending them to a better climate was worth considering, together with the almost certainty that personal reclamation would be more likely to follow in a land which gave better opportunity. After two years or so of deliberation the Pitt Ministry determined on carrying out the joint experiment.

It should be stated, that Matra had no part in these things, after all. Through Banks's influence he obtained a secretaryship in the British Embassy at Constantinople; and afterward a consular appointment at Tangier, which he held for over twenty years. He wrote long letters to Sir Joseph, full of interest to him in botany, and in Moorish travel, especially with reference to the exploits of Hornemann and Park and Lucas. He makes a complaint, the like of which has been heard since, that the English Government treats its Consuls with great indifference. Coins from Barbary were duly gathered for Miss Banks. Stories of Moorish superstition and craftiness fill many entertaining pages. And Sir Joseph rewarded him with return letters, which he well deserved.

The first fleet for Botany Bay was at length dispatched in May, 1787. It included six transports, three storeships, H.M.S. *Sirius*, and the *Supply*, tender; all under the command of Captain Arthur Phillip. They carried more than one thousand persons, including a number of male and female convicts, and a first party of settlers. Their destination was reached safely in January, 1788. After some consideration, Phillip determined to make the settlement at Port Jackson, as presenting far the most suitable conditions for the purpose. It happened that two ships were standing in that famous harbour, under the command of the gallant La Pérouse. The English captain resolved to lose no time in selecting a station,

which he called Sydney Cove, after the then chief of the
Colonial Department.   Thus it was maintained after-
ward, that Captain Phillip had forestalled a possible
French occupation by only six days.

How the new colonists fared under Governor Phillip :
in settling and planting ; in travels of discovery ;
in troubles with the convicts ; in periods of scarcity of
food : are matters of high romance which belong to the
early history of Australia.   Unaccountably, the historians
of England have left these stirring occurrences severely
alone.[1]

In all these things it would appear that Sir Joseph
Banks had a consultative share.   Indeed, there is reason
to believe that he actually refused high office, on the
ground that his services would be of more value to
the Colony if he kept aloof from ministerial responsibilities.

A memorandum in his handwriting runs thus :   " I
could not take office and do my duty to the Colony.   My
successor would naturally oppose my wishes.   I prefer,
therefore, to be friendly with both sides." [2]

But there were other matters within his entire personal
control.   A new energy in exploration was aroused.   A
long period set in, during which Banks was watching the
products of the country and sending out naturalists at
his own expense.   The first botanical results came into his
hands before the close of the year 1789.   The second and
following fleets took out selections of seeds for the use of
the colonists, and bore new instructions for the careful
handling of plants destined for Kew.   Moreover, he had
an estate of his own ; a memorandum of 1790 gives instructions as to the supply of each family with ten convict servants ; victuals for three years ; cows, swine,

---

[1] Except Mr. Lecky, who dismisses it all in a few lines.
[2] *v. Hist. Records of N.S. Wales*, vol. I, pt. 2, p. 229.

sheep, seed, corn, etc., for two years ; together with tools and necessaries.[1]

Captain Phillip remained as Governor of New South Wales until relieved by his comrade Hunter, in 1792. Phillip developed high qualities in his post, and succeeded in winning general approval for the manner in which he had carried through its initial difficulties. His task was by no means a light one. As might have been counted upon sooner or later, a mutiny broke out among the convicts. The arrangements for provisioning the people were by no means adequate to their need, and more than once they were on the verge of famine. The first of the ships sent out with fresh supplies was wrecked, and everything on board lost. The next one brought two hundred and twenty-two female convicts, many of them invalid and decrepit ; with hardly enough food for the daily rations of the people. This, at the very time when the Governor was advising, for the time, some restriction in dispatching convicts, excepting carpenters, masons, bricklayers, and farmers. The reproductive powers of the country were hardly yet in fair activity. These shortcomings, added to those resulting from the large proportion of ignorant and helpless persons in the Colony, made Phillip's task particularly arduous. Yet he triumphed in great measure over these adverse circumstances, and handed his authority to Governor Hunter with good promise for future years.

Hunter's period of office was a time of increasing prosperity. As far as concerned individual conditions, everything was improving for the colonists. His successor, in the year 1800, was Philip Gidley King. It is remarkable that these three officers all proved such able administrators, and reflected such honour on the Naval Service. Writing to Governor Hunter in March, 1797, Banks is very enthusiastic over the prospects of New

[1] *Op. cit.*, p. 424.

South Wales. He wishes he could go himself, and settle
on the Hawkesbury River :

" I see the future prospect of empires and dominions
which cannot be disappointed. Who knows but that
England may revive in New South Wales when it has
sunk in Europe ? "

Plans for exploration were thought of. It was intended, in 1798, to send out Mungo Park in charge of an
expedition to the interior. At present, life was centred
in the little settlements at Port Jackson, and on the
Hawkesbury and the Paramatta rivers. As yet there
was no evidence that the Colony would furnish a return
in merchandise for the expense and the exertions of the
Mother Country. In truth, the steps hitherto taken had
been too rudimentary. For several years the settlers had
been dependent for food and necessaries upon importations from England. And the penal side of the population made a horrible drag on the resources of the little
colony.

Several men were distinguishable from the crowd of
settlers by their intelligence and energy. One of these
was John MacArthur. He had come out with Major
Grose, whose mission was to raise a New South Wales
Corps. Lands were allotted to the officers of this body ;
a step which more than anything else contributed to the
prosperity of the Colony. Indeed, Grose declared later
that these were the only settlers to be relied upon.

In 1794 MacArthur had two hundred and fifty acres
in cultivation. As a practical farmer, he left his brother-
officers behind ; and was presently given (by Grose) a
sort of Inspectorship of the district. At length, several
sheep brought from the Cape of Good Hope were offered
for sale. They were of a Spanish breed, which had been
thriving at the Cape, and MacArthur bought some of
them. After several years' trial as a breeder, he met with
surprising success. In October, 1800, he sent eight

fleeces to London, for inspection and opinion. Meanwhile, his flocks and his acres increased " by leaps and bounds."

In 1803 MacArthur was sent to England under arrest, because of a duel in which he was concerned. As it happened, he was the very man wanted in London at the time. Manufacturers in wool all over the country had got wind of the soft and beautiful wool that had come from New South Wales. They sent Memorial after Memorial to the Office for Trade and Plantations. But there was one important person who was indisposed to see any particular merit in this wool : Sir Joseph Banks. Mr. Fawkener wrote from the Office making him acquainted with the Memorials, and asking his sentiments on the subject ; for Captain MacArthur represented to the Ministry that he had found " from an experience of many years that the climate of New South Wales is peculiarly adapted to the increase of fine-woolled sheep, and that from the unlimited extent of luxuriant pastures with which that country abounds, millions of those valuable animals may be raised in a few years with but little expense than the hire of a few shepherds." Banks replied that he had seen some fleeces from New South Wales, but none equal to the best piles of old Spain. He had no reason to believe that the climate of Australia was better calculated for the production of fine wool than that of the temperate climates. " I am confident that the natural grass of the country is tall, coarse, and reedy, and very different from the short and sweet mountain grass of Europe, upon which sheep thrive to the best advantage. . . . I confess I have my fears that the Captain has been too sanguine in his wishes to give a favourable report of the country, and that it will be found on enquiry that sheep do not prosper with them unless in lands that have been prepared for their reception with some labour and expense."

It was a long, and it must be said lazy-minded, letter,

bearing some signs of failing powers : as though his
now frequent attacks of illness impaired Banks's efforts
to give the utmost attention to matters before him. All
this was nothing to MacArthur. There were four thousand sheep on his farms. He expected them to double
every two and a half years. Ten years of experience
lay behind him. Telling his own story at the Office of
Trade, he convinced the Committee that there was a
great future in store for New South Wales. The statistics
which he offered would show them the possibility of
supplying England with any quantity of the finest wool
that she required. And he was determined, on his return
home, to devote his entire attention to the subject. The
support of persons interested in the matter soon became
highly encouraging. He went home having apparently
converted everybody. The Government made him a new
grant of ten thousand acres. He took back with him five
Spanish merino sheep which he had bought at the King's
sale in 1804,[1] and a number of plants and fruit trees.
He now became the leading planter and agriculturist
in the Colony, extending his operations constantly, and
adding largely to its prosperity. Hence MacArthur came
to be called the real Founder of New South Wales.

A friendly competitor was the Rev. Samuel Marsden,
chaplain and magistrate. Governor King describes him
to Banks as the best practical farmer in the country.
In 1804 he owned twelve hundred sheep, and held seventeen hundred and twenty acres. He wrote a few letters
to Banks with encouraging views of sheep-breeding, and
proposed experimenting with Leicester and Lincoln
sheep, two rams of which he had instructed his agent
to export.

---

[1] The prosperity of the King's farming is well shown by the prices
given by MacArthur : £6 15s. od., £11, £16 16s. od., £23 2s. od., £21,
£28 7s. od. The heaviest fleece weighed 7 lb. 2 oz. ! Banks also gave
some heavy prices at this sale.

Yet these people could not stand Prosperity. So many petty conflicting interests came into being, that quarrels, and efforts at supremacy, and even lawlessness, arose, that the peace was broken over and over again. Captain MacArthur, for example, defied the laws in a very discreditable way, and caused a deal of trouble in the Colony.

When Governor Hunter's term of office was ripening, and the choice of his successor impended, the authorities at home decided upon sending Captain Bligh. This was on the warm recommendation of Sir Joseph Banks. Bligh was disappointed to learn that the Governorship would be separated from the Naval service in future. His services during the war were very honourable. He had quelled several mutinous attempts when on cruise. " Duncan knew the value of your conduct in action, though he did not choose to praise it ; and Lord Nelson not only knows how you conducted your ship when it lay alongside of his Lordship's at Copenhagen, but will not omit any opportunity of giving you credit." [1] And he had hoped to reach the high ranks of the Service. Banks assured him he would not have advised him to take a step " which may I fear on a future occasion be interpreted by the Admiralty into a dereliction of your chance of a Flag."

However, Bligh was sent to New South Wales. His views and those of Hunter coincided, and he tried his best to fulfil the latter's wishes. As is not uncommon, the drink question was prominent ; and existing regulations forbade the distillation of spirits. But MacArthur must needs import a still from England. For this, and for using seditious language, he was brought to trial. Bligh, up to this moment, had been generally popular. But when he began to insist upon the law being adhered to, he at once raised up a number of vindictive enemies.

[1] Banks to Bligh, September 17, 1805.

Party spirit was fierce, and MacArthur was acquitted. The Colony was a volcano in active operation. At length, affairs culminated in the arrest and imprisonment of Governor Bligh, by Major George Johnston, one of the disaffected persons who were profiting by the trade in ardent spirits. Johnston was presently sent home to be court-martialed, and duly cashiered. This was very damaging to Governor Bligh, who had only done his plain duty, and with as much consideration as possible for offenders. The Government at home were obliged to recall him, and he returned to England. He was made a Vice-Admiral in 1814.

Colonel William Paterson remained in command: a man very popular, who had been Lieutenant-Governor under King. He was of the New South Wales Corps, and made a good settler ; and was, besides, an indefatigable Botanist.

Sir Joseph Banks to Mrs. Bligh.

" April 16, 1809. . . . The Government have not altered their intention of recalling your husband. Good often arises out of evil. His conduct, always in my opinion honourable, just, and equitable, has been more than once canvassed in his absence, where he has been unjustly, indecently, and scandalously misused. Unfortunate as we are in the unwise decision of His Majesty's Ministers to recall him, this evil must produce the good consequence of his being able to answer for himself such charges as may be brought against him. The result, you and I cannot doubt, will be favourable ; and will throw a light on former transactions, not a little important to his good fame."

All these things, however, thrust into Banks's hands because high officials found him an indispensable adviser, were not the affairs which gave him most concern. He was a Botanist before everything, and, while New South Wales was struggling from the condition of a much-handicapped settlement into that of a prosperous agricultural Colony, his eye was eagerly and constantly turned toward the indigenous Flora of the country. This is, in some respects, one of the most important of Banks's utilitarian efforts ; for it opened a profoundly important era in the advance of Botanical Science.

Few things delighted Banks more than finding a new Botanist. One day in March, 1795, he received a letter from a distant admirer at Manchester, who wished to gain employment as botanist. He knew something of farming, and of horse-keeping, but wished to be more in touch with the study of Nature. He sent a *Drosera*, and one or two other items of his own gathering. It was the letter of a person in humble life, who was getting a sort of education through books and observation. Banks, with unfailing good-nature, promptly replied.

Sir Joseph Banks to George Caley.

(March, 1795.) " SIR,—I do not know there is any trade by which less money has been got than by that of Botany. . . . If you wish to apply yourself to the study, the only means I know of raising yourself into notice is to learn the names of the curious plants cultivated in the Botanic Gardens. English Botany is well known to many persons in the country, who practise it as an amusement. Nothing, therefore, is likely to be got except in the exotic line ; and in that, unless you are both diligent and fortunate, you will not succeed. If you have bodily health and strength, and understand the business of a gardener's labourer : that is, I believe, the only good method of getting instruction. We have several foreigners who every year enter in that capacity in the King's

gardens, and some of them persons of property. They
receive about ten shillings a week, upon which they can
maintain themselves ; and if they behave they have
great opportunities, not only of studying the culture of
plants but also Exotic Botany.

" If you resolve to take this step, I will undertake to
recommend you to some good garden here ; and if you
continue diligent and make a proper progress I will, from
time to time, give you such assistance as will make your
station less disagreeable than otherwise it would be.

" The *Drosera* you sent is the real *longifolia* of Linnæus,
Hudson, and Withering. . . . The Moss you have sent
appears to me a new one, and the finding it does you
great credit. It is, however, so young that it is impossible
to determine with certainty to what genus it belongs.
I shall be obliged to you if you will gather some more of
it and send it to me by the mail coach in a little parcel
wrapped up in wet moss."

Such was the beginning of an acquaintance which
lasted during Banks's life. Caley came up to London,
and was presently employed at Kew, at the Brompton
Botanic Garden, and elsewhere. He had his difficulties,
arising mostly from an irregular temper, and an ill-
mannered way of looking at disappointment or misunderstanding. Banks was very generous in occasional help,
but was forced to speak sharply to him sometimes in reply
to what was downright impertinence. When the New
Holland Exploration scheme was proposed, Caley lost
the opportunity of an appointment, through having recently left Kew in a fit of discontent at his " poor pay and
bad prospects " ; and he was almost offensive with his
veiled reproaches. It is surprising how Banks endured
it all. He wrote (August 27, 1798) : " You seem to have
so good an opinion of the value of your own abilities as
to think that I ought to demand of Government to send

Q

you out. Suppose I was to agree to this opinion of yours, and a competent salary was to be allotted to you, pray tell me what you would propose to do for your employers in return. . . . Let me know your opinion on this head, and I shall be better able to judge what my conduct to you in future ought to be." Other passages in this letter show that Banks was exercising considerable forbearance toward him. There was probably no further immediate communication between them. But, in the ensuing November, Sir Joseph wrote to tell Caley an unexpected opportunity had arisen for him. Governor King was going back in the *Porpoise*, and offered to take him if he chose to go. Banks was willing to allow fifteen shillings a week, and would not exact further terms but to be supplied with specimens of new and curious plants. He enclosed a five-pound note to come up to London with, and promised to advance something toward an outfit. After long delay, Caley reached New South Wales. He proved a capable botanist, and an indefatigable collector. He soon sent home substantial proofs of his industry. As time went on he became a prosperous settler, with a farm and garden on the Paramatta ; and took some share in public affairs. He warmly supported Governor Bligh, and tried to sway local opinion in his favour. According to Caley, Bligh was well-meaning, but tactless ; unable to control the evil passions certain to exist in a young colony. His ability was great in everything except to combat open rebellion.

Caley's long letters are full of material for the early history of the settlement. Many of them are printed by the Sydney Government in their *Historical Records*. The botanical portions are the work of a thorough enthusiast in the science. Some of his consignments are prodigious, as in August, 1804 : sixty skins of birds, two of kangaroo, forty papers of seeds, and two hundred and twelve pages of description ; in April, 1805, forty-one species of seeds, and three hundred and eighteen pages of description. As early as 1805 he was thinking of returning to England. Three years later there is a growing discontent with existence ; partly because Bligh was not interested in Discovery, and he had no other neighbour to confer with concerning plants and Natural History ; and partly through local politics. He lives like a hermit. His neighbours are contemptuous over his occupation. Yet there is a silver lining : he is grateful that it is " a country which gives him an opportunity for distinguishing himself." Beside this, it is evident that reflection and reading were making a man of him. His pages become enlivened with curious anecdotes and shrewd observations. And an occasional adventure in the forest or the desert reveals the dangers and the difficulties of his pursuit.

After a long interval, a letter came from Sir Joseph. It was no response to Caley's sad-coloured tales, and his hundreds of pages. The decline in Banks's health, and the number of leading-strings he continued to hold, made it increasingly difficult to do justice to all the interests in his charge. This letter is from a man beginning to recognize that the burden of life was telling upon him, and he would gladly relinquish some of his self-imposed responsibilities.

Sir Joseph Banks to George Caley.

" SOHO SQUARE, *August* 25, 1808.

" I have been a long time prevented from writing to you by increasing age and infirmities, principally by having the gout upon me with severity at the times when opportunity of letters offered.

" You have in general been an active, a diligent, and a useful assistant to me in your present situation ; and I have found you on many occasions to possess a strong understanding. I cannot, however, agree with you in the propriety of your having refused to deliver up the plants entrusted to your care by Mr. Brown, when Governor King came home . . . his return was a good opportunity, for many of the plants he brought home came safe and in good condition. . . .

" I have grown of late years very infirm ; my eyes fail me very much ; and I have not, of course, the pleasure I used to have in the pursuits of Natural History. I have not, therefore, any longer occasion for your services in the extensive manner in which you have employed yourself of collecting great quantities of articles. You deserve, however, some reward from me for your diligence and activity.

" You have, I understand, the lease of a farm from Governor King. If you wish to employ yourself in the cultivation of it, or if you wish to return home, I am willing to settle £50 a year upon you for your life, and to release you from all services to me beyond what you voluntarily wish to perform. You would probably choose, if anything new should fall in your way, to send it to me. But as I mean your annuity as a recompense for past services, I shall not bind you to any future ones till I hear from you on this subject ; and, till the whole can be arranged and settled, everything to go on as it has hitherto done.

" Mr. Brown and Mr. Bauer are well. They are busily employed in arranging and making drawings of the immense collections they have brought home. . . . Your specimens and descriptions are carefully preserved for you."

Caley came back to England in 1811. Three years later he went out as superintendent of the garden at St. Vincent's. He was not a social success, and did not get on with his neighbours. Finding that seeds, etc., were stolen from the garden by Sunday visitors, he withdrew permission for the public to enter on that day.

This was justifiable under the circumstances, but inexpedient ; and an occurrence of this kind naturally added to the local prejudice against him. Caley was a good Collector. Numerous plants and seeds were sent to Kew. He stayed at St. Vincent's till 1819, when he came home ; and spent much of his retirement in Kew Gardens.

Another botanist who went out to New South Wales at Sir Joseph Banks's instance was George Suttor. He appears to have gone purely as a settler in the country ; taking with him a variety of fruit trees, and promising to send home to Banks what new plants he could find. He sailed in the *Porpoise*, in November, 1798. He turned out a very able colonist, cultivated the vine, and was doubtless the originator of the wine produce of Australia. He visited England in 1842, became a F.L.S., and published a paper on the Culture of the Vine and the Orange in Australia. He died on his estate at Bathurst, in 1859.

Suttor wrote a short life of Sir Joseph Banks,[1] marked by the warmest sentiments of regard and esteem.

[1] London, 1855.

CHAPTER XIV

CAPTAIN FLINDERS AND ROBERT BROWN

AMONG the sailor-adventurers of this period, few men deserve more honour than Matthew Flinders. Born about 1760, he served some time in the mercantile marine before entering His Majesty's Navy. In 1795 he was on board a ship carrying out a Governor to Botany Bay. George Bass, the surgeon, found a congenial mind in Flinders. Each wanted to explore something; and after their arrival in New Holland, they made important surveys and discoveries in company. They were presently sent out on a trip southwards, in order to settle the question whether Van Diemen's Land was an island or no. They found open water, which was forthwith called Bass's Straits. They were afterward engaged on the still more important task of defining and perfecting Cook's researches on the east coast of New Holland. Then Flinders came home, and was deservedly promoted. Bass presently went to sea again, in another direction.[1]

In 1800, the *Investigator*, 334 tons, was ordered out for a scientific voyage to New Holland, under the command of Flinders. An astronomer, a botanist, two draughtsmen, a gardener, and a miner were chosen to go with this Expedition. William Westall, afterward a famous landscape painter, was the draughtsman. Ferdinand Bauer, the botanic draughtsman; Peter Good, the

---

[1] According to Jörgensen, Bass was entrapped ashore in Chili, and ended as a captive in the silver mines.

gardener, had been in charge of plants between England
and India ; John Allen, the miner, appears to have come
from Derbyshire, where Banks had some mining interests ; John Crosley was the astronomer. The naturalist
was Robert Brown, who had been fortunately brought
to Sir Joseph s notice by his friend Correa de Serra,[1] in
the following message to him while away from London :

"SOHO SQUARE, *October* 17, 1798.

". . . Mr. Brown, a very good naturalist, who fre-
quents your library, where I have made acquaintance with
him, hearing that Mungo Park does not intend to go to
New Holland, offers to go in his place. Science is the
gainer in the change of man, Mr. Brown being a professed
naturalist. He is a Scotchman, fitted to pursue an object
with constancy and a cold mind. His present situation
is of Ensign and Assistant Surgeon in the Fifeshire
Fencibles, previous to which employment he received
a regular literary education at Edinburgh. It is by his
own desire that I take the liberty of making you acquainted with his wishes ; his modesty debarring him
from writing to you himself. . . ."

On making acquaintance with Brown, Banks found a
true disciple. The current of the young man's life was
eventually turned to the single pursuit of Science ; and a
close friendship existed between them as long as Banks
lived.

In December, 1800, Ensign Brown was offered the post
of naturalist on board the *Investigator*, at a salary of
£400. Sir Joseph writes : " If you choose to accept the

---

[1] Josef Correa de Serra was a Portuguese exile in London, whose
tastes were so entirely those of Banks that an intimate friendship
arose between them. An interesting account of an examination of the
Lincolnshire coast made by them in company, in 1796, was included in
the *Philosophical Transactions*.

appointment I will certainly recommend you. . . . I will not mention any other person till I have heard from you, and hope that you will be the messenger of your answer." He obtained leave of absence, and gladly accepted the new commission. Captain Flinders sailed from Spithead on July 18, 1801. The ship called at Madeira, and reached the Cape of Good Hope in October. From this place Flinders wrote that Mr. Crosley's health was breaking and he had decided to quit the Expedition and return to England. Captain Flinders replies, in advance, to the question whether another astronomer should be sent to him, by promising that he and his brother (a Lieutenant) could perform the whole business of the Observatory; and the latter would leave it to the Board of Longitude to decide as to his remuneration.

The next letters from the *Investigator* are dated from Port Jackson, in May, 1802. The naturalists were already active. Mr. Brown sent home some seeds, and reported that seven hundred and fifty new plants had been observed. Mr. Bauer had made many drawings of plants and of animals, and was indefatigable in studying the fructification of plants. Mr. Good was a " most valuable assistant." Brown mentions the want of a garden on board ship, and it would most likely be prepared. Beside this long letter to Sir Joseph, he also writes to Dr. Dryander, and to that ardent horticulturist, Mr. Charles Greville.

A letter from Peter Good, dated July 20, 1802, states that the ship is in harbour in Port Jackson. Good tells of his excursions, sometimes with Mr. Brown and party, and sometimes by himself. He has added above five hundred species of seeds to the former collections. He has taken lodgings on shore for the convenience of drying specimens, and has spared no labour nor expense in the performance of his duties. " We sail from here to-morrow on a long cruise with the *Lady Nelson* in company.

The garden has been put on board some days ago. . . . The two French ships on discovery are now in the harbour. The *Géographe*, which we met with on the south coast, came in here a few weeks ago in distress ; almost the whole crew sick, and in great want of provisions and water. . . ."

The appearance in these seas of the *Géographe* should here be accounted for. Flinders's fate was ultimately affected by the circumstance that a French voyage of discovery was in hand, under the command of Captain Baudin. Flinders had met the French ship in Bass's Straits. There was friendly communication between the two parties ; but reasons existed for suspecting that the French meant, if possible, to forestall the Englishmen in the possession of some part of New Holland. In later years this proved to be the case. Baudin's *Découvertes*, as published after his return home, included hundreds of leagues of coast-line, already named by Flinders and Bass, now appearing with French titles.

The *Investigator* now proceeded on a cruise, and surveyed the great and terrible Barrier Reef, and the Gulf of Carpentaria. Then they stayed at Timor. From this island dispatches were sent homeward, the purport of which included the rotten condition of the ship, and slackness in the Natural History department. Peter Good suffered from dysentery on the return voyage, and died two days after reaching Botany Bay. This was a serious loss to the Expedition. Mr. Brown testifies to the value of his indefatigable exertions in his department. The garden on board ship now contained upwards of a hundred growing plants, and these must needs suffer from the loss of Good's care and zeal. Moreover, it had to be transferred to the *Porpoise*, the vessel secured by Flinders for the homeward voyage.

Scientific observations were now resumed at Port Jackson. Beside the Botany and Zoology, a record was taken in Astronomy and Meteorology. A small observatory was set up, which was mostly in the charge of John Franklin, a young midshipman who was already distinguishing himself in nautical and surveying work. Franklin was usually selected for a companion when Captain Flinders made excursions of any sort for scientific purposes.

These were the days of Governor King at Port Jackson, a man who was of great assistance to the party in many ways. And they soon had need of one who would help them, and could do so. It was at length determined that Captain Flinders should go home in the store-ship *Porpoise*, while Mr. Brown and Mr. Bauer remained in the country in order to proceed with their Natural History studies. The Governor undertook to make them comfortable, provide them with a house and one or two convict servants, and a Government passage to England in due time.

About this date arrived a letter from Sir Joseph, in a tone which must have given some cheer to the little party.

Sir Joseph Banks to Robert Brown.

" Soho Square, *April* 8, 1803.

" It gave me sincere pleasure to learn by yours from Port Jackson that you was in health, and had been so fortunate in the first part of the interesting business to which you have so handsomely volunteered yourself.

" Your commander deserves, in my opinion, great credit for the pains he must have taken to give you a variety of opportunities of landing and botanizing. Had Cook paid the same attention to the Naturalists, we should have done more at that time. However, the bias of the public mind had not so decidedly marked Natural History for a favourite pursuit as it now has. Cook might have met with reproof for sacrificing a day's fair wind to the

accommodation of the Naturalists. Captain Flinders will
meet with thanks and praise, for every sacrifice he makes
to the improvement of natural knowledge which is com-
patible with the execution of his orders.

" The seeds you sent by the ship that brought your
letters came safe and in good order. They are all sown
in Kew Gardens, and much hope is built on their success,
which will create a new epoch in the prosperity of that
magnificent establishment by the introduction of so large
a number of new plants. . . .

" I very much approve of your employing yourself
when in harbour, rather in making descriptions and
enlarging your observations than in attempting to make
copies or to prepare anything to be sent home. Without
you the specimens or descriptions you might dispatch
would lose much of their value. . . ."

Captain Flinders sailed homeward in the *Porpoise* in
company with the *Bridgewater* and the *Cato*, two small
trading vessels. They took the eastern side of New
Holland. All went well for a few days, and then disaster
fell upon them. The *Porpoise* and the *Cato* struck upon a
coral reef, and speedily went to pieces. The precious
garden was lost. Most of the ships' company were left
under temporary shelter upon a sand-bank, while Captain
Flinders returned to Port Jackson in the *Porpoise's*
cutter. Westall, and Allen the miner, were also bound
for home, and were among the stranded sailors on the
reef.[1] During his absence of about six weeks, the ship-
wrecked party sowed the seeds of maize, oats, pumpkins,
etc. The young plants were coming up when the schooner
Cumberland hove in sight, with Flinders on board. In
company with this vessel was a trader bound to China.
Westall and young Franklin elected to go home via

---

[1] Among Westall's fine illustrations to Flinders's published story,
there is a good picture of this scene.

Canton. Some others returned to Port Jackson, and the
most of his own crew resolved to share Flinders's fortunes
in the schooner. The vessel was hardly fit for a long
voyage. But Flinders was determined to get home to
England, report himself and his adventures, and come
back to New Holland with better equipment, and resume
his surveys. The *Cumberland* sailed round the north of
Australia, touched at Timor, and then headed for Mauritius, at that time in the possession of the French. She
anchored at Port Louis on December 17, 1803.

Flinders was in possession of a " pass," in the interests
of Science, by which he was understood to be secure from
any hostility on the part of French vessels of war. Baudin
was provided with a protection as against the armed ships
of England. A similar civility was exercised on other
occasions, and the terms of a pass were usually respected.
But Captain Flinders, reaching Mauritius in an unsea-
worthy ship, with no thought of missing the friendly rites
of hospitality, was seized on pretence of his being a spy.
His ship was confiscated, together with his papers and all
the fruits of his surveying toil. Whether he was detained
as a suspected person, or as a naval officer in arms against
France, was never made clear. De Caen, the Governor,
is represented as an illiterate person, and might be
acting only in the manner of a Jack-in-office, a character
not unfamiliar to us in the story of French colonization.
Yet, the fact of Baudin's story being published with his
claims of wide discovery, his *Terre Napoleon*, etc., in the
end satisfied most people that De Caen was alive to the
opportunity of seizing any possible occasion for hamper-
ing English exploration.[1]

Captain Flinders was treated with some severity at

[1] This notion is supported by the fact that copies of Flinders's chart
of Van Diemen's Land were struck off for Baudin's use (*Philos. Mag.*,
XVI, 265) ; and by Flinders's remark that he saw the *Cumberland* used
at Port Louis, and her stores consumed. A statement that the ship was
afterwards in service under Baudin we have not been able to verify.

first, until upon medical advice he was given a lodging in the town under close supervision. Application was made in vain for permission to use his journals and charts. Even a " spy-glass " was regarded as a legitimate forfeit to his captors. Matters were a little improved by the captain being confined in the Garden Prison, where other English officers were under detention, and who, by the way, had also lost their spy-glasses. The first news of all this reached London in August, 1804.

Captain Flinders suffered a good deal from depression of mind. But he kept active, and spent much time in noting the productions of the island and storing up material for his book of travel. Then his health improved. It was even a matter of discussion with his French friends (of whom he had many) as to whether he should not reconcile himself to his fate and settle in the island. However, he would have none of this. He was determined on restitution. He wanted promotion in the British navy, and he would return to the wife he had married a few weeks before leaving England. Thus year after year rolled on. In July, 1809, Sir Joseph Banks had a letter from Flinders (dated September, 1807) stating that he had got back his books and papers. An order had come to set him at liberty. This was after numberless applications, remonstrances, promises, delays, during which he had seen ship after ship arrive and depart, ever awakening hopes that could not be realized. Flinders's difficulty now was, mainly, that he kept up a high tone as a British officer ; and this De Caen affected to disbelieve. Even after repeated orders to release his prisoner the Governor persisted in regarding them as open to his discretion. Doubtless personal pique on both sides delayed relief.

It was not until June, 1810, that Captain Flinders found himself on board the *Otter*, Captain Tomkinson, bound for the Cape. He reached Spithead on October 23.

He was very well received at the Admiralty. Croker, and Barrow, and Yorke showed a disposition to appreciate his misfortunes, and to consider that he had creditably added to the annals of the British sailor. Banks was very cordial to " his countryman," as he called him—for Flinders came from Lincolnshire. He died in 1814, his impaired constitution lasting only long enough to permit him to finish writing the story of his *Voyage to Terra Australis*.

Several years elapsed before the botanists heard anything of their unlucky leader.

Meanwhile, affairs at Port Jackson went on with tranquillity, tempered by zeal in the study of Natural History. George Caley joined the party, as suggested by Sir Joseph Banks. The loss of the previous collections was made good, and an immense number of plants, etc., were made ready for dispatch to England. Governor King visited their garden whenever the botanists were absent collecting, and entered heartily into all their plans for safely transporting them across the ocean. As far as plants and seeds were concerned, most of the work had to be done anew, on account of the loss of the *Porpoise*.

Nearly two years elapsed before the decision was made to go home. After much difficulty in finding a suitable ship, it was determined to utilize the old *Investigator*. Governor King found that though her upper works were practically rotten, the hull was tight and seaworthy. And Brown and Bauer might go with their collections safely and successfully. The remaining garden was left in Caley's hands, who proposed to take to England his collections by a later opportunity.

Mr. King was careful to provide a safe conduct for the ship as far as he could, in a letter to the " Prefect or Officer, civil or military, commanding in any port belonging to the French Republic where H.B.M.'s ship the *Investigator* may eventually be carried," with a view of saving at least the scientific collections. She sailed in May, 1805, and met with no trouble beyond the inevitable tedium of five months unrestingly spent upon the sea. Mr. Brown wrote from Liverpool, October 13, a long letter to Sir Joseph Banks, announcing his arrival.

The collections on board had been cared for, and the Captain took a share in the efforts to preserve them. But they were beginning to suffer from the wet state of the ship. Brown was indisposed to let them remain for the voyage round to London, and sent them all up to London by road. We have particulars of the enormous cost of this operation. Cartage amounted to £53. Unloading and reloading at the London Custom House, insurance, fees, etc., brought the bill up to £87 8s.

Sir Joseph was not able to take a personal share in the botanist's affairs and his triumphant return. Just at this moment he was laid up with a bad attack of gout, which, spite of abstemious habits, now seized upon him with greater frequency than ever.

Sir Joseph Banks to William Marsden.

" SOHO SQUARE, *January,* 1806.

". . . After tedious delay, the unavoidable consequence of a severe fit of illness which has given me much mortification, I am at last able to fulfil my promise of giving you some account of the nature of the collections made by Messrs. Brown and Bauer, in the course of the voyage of discovery from which they have lately returned, and of suggesting such measures for the consideration of their Lordships as are most likely to secure to the public the fruits of the labours of these very active and industrious men. . . .

" An exact account of Mr. Brown's collections cannot be given till the examination and comparison with books, which employs his time at present, has been finished.

In the meantime, they consist, as far as he is able to judge, of the following particulars.

" Specimens of plants—

From the S. coast of New Holland .	700 species
From the E. ,, ,, .	500 ,,
From the N. ,, ,, .	500 ,,
From Port Jackson and neighbourhood	1000 ,,
From Van Diemen's Land . . .	700 ,,
From Timor	200 ,,
Total plants . . .	3600 species
Dried skins of birds	about 150
Insects	one case
Minerals	three boxes

" The arrangement of these Birds, Insects, and Minerals he wishes to transfer to persons more conversant in these branches of Natural History than he considers himself to be.   The plants . . . he wishes himself to arrange and describe if their Lordships shall be pleased to employ him in that duty.

" Mr. Ferdinand Bauer and his brother, who have the honour to be Botanical painters to His Majesty, at Kew, are nearly equal in ability.   They are allowed to be the most skilful painters of Natural History in the kingdom, and in my poor opinion are not equal'd in Europe.   Mr. Ferdinand Bauer was induced to undertake the voyage by the generous encouragement held out by their Lordships when the voyage was first planned. As a sample of his skill I have the honour to send with this a drawing made by him of a very curious and inter-esting plant. . . .   The quantity of sketches he has made during the voyage, and prepared in such a manner by references to a table of colours as to enable him to finish items at his leisure with perfect accuracy, is beyond what I thought it possible to perform. . . .

"Sketches of plants made on the coasts of
New Holland and New South Wales . . 1541
Sketches of plants made on Norfolk Island . 80
Sketches of plants made at Timor . . 60
,, ,, Cape of Good Hope 89
Sketches of animals on Norfolk Island . . 40
,, ,, in New Holland . . 263

Total 2064

". . . I beg leave humbly to suggest to their Lordships that the salaries of Mr. Brown and Mr. Bauer be continued to these gentlemen for such time as their Lordships shall think proper. . . . I will undertake to direct the progress of these gentlemen, to quicken them if they are dilatory, to assist them when it is in my power, and to report to their Lordships the progress made by each in his respective department once a year at least, or oftener if required to do so. . . ."

Sir Joseph further suggested that orders be given for all their treasures to be deposited in the British Museum, and finally arranged there by competent officers. Also that Messrs. Brown and Bauer should be recommended to join together in publishing engravings and descriptions of their most interesting objects in Natural History, in a handsome form, with a view to their own profit.

Mr. Marsden, on behalf of the Board of Admiralty, wrote in reply, conceding all Banks's suggestions, and promising to bear the expense of carrying the collection from Liverpool to London. It would appear that some of Banks's enthusiasm had reached their Lordships. It certainly was a magnificent gift to the Nation. They could not fail to see the accumulating grandeur of Sir Joseph's plans. Again and again he had generously

R

considered the public good, and devoted his resources toward defraying the cost.

Banks now invited the Lords of the Admiralty to regard his projects with some consideration ; and professed himself quite willing to modify the terms of his proposed gift in any way which might be expedient in the case of their having views which differed from his own. He supposed that the operation of arranging the plants in systematic order would occupy at least three years, and that a great number of the paintings would be finished in that time. The whole Collection was at present housed in Soho Square.

And now began the task, the full performance of which carried Robert Brown to the highest distinction as a botanist. He was occupied for four months in a first arrangement of his plants according to the received methods of the time. With the assistance of Banks and Dryander, a process of selection was undertaken with a view to preparing specimens suitable for the Museum. Several years of assiduous labour passed in a more thorough examination, from which Brown emerged the founder of a better classification than had yet proved finally acceptable. He recognized the value of Antoine Jussieu, and loyally followed his steps. He saw, further, the import of Goethe's suggestions that the Leaf is the common type of all the varying organs of the Plant. The adoption of new views was confirmed in the pains he took with his immense mass of material.

The first published fruit of Robert Brown's labours was *Prodromus Floræ Novæ Hollandiæ et Insulæ Van Diemen* (London, 1810). This work was received with acclamation by all the botanists of Europe. Congratulations poured in from France, Sweden, Germany, Italy. Alexander von Humboldt hailed him as *Botanicorum facile princeps;* and that title adheres to his name to our own days. He deserved his fame, for Brown was the

H. W. Pickersgill, R.A.

ROBERT BROWN, F.L.S.
From a picture in the possession of the Linnean Society

true man of genius, in whom imaginative power is wedded to alert habits of observation. Lonely plant-hunting in the wilds ; accumulating an enormous number of examples ; following with a careful arrangement according to existing authorities : these things might be undertaken by many an ordinary mortal, gifted with the necessary habits of intelligence and industry. Genius once more went in advance of the ordinary mortal. And Genius, as is its wont sometimes, woke up to find itself famous.

Brown became librarian to Sir Joseph Banks, after the death of Dryander in 1810. Henceforth he is an inmate of the house in Soho Square, and his public life is largely mingled with the remaining career of Sir Joseph. They were close friends until the end. Brown had a remarkably gentle and lovable disposition, and was blessed with troops of friends, who adhered to him through a long and useful life. This ended as late as 1858, so that there are veterans still living who were among his associates.

Banks's library and collections were transferred to the British Museum in 1827, Mr. Brown being appointed keeper.

CHAPTER XV

ICELANDIC AFFAIRS

THE French revolutionary war was answerable for many unexpected side issues. Not a small one was the idea of annexing Iceland and the Faroes to the British Crown. And it would have been done but for the normal restraint of English statesmen. For once, Sir Joseph Banks was a politician, and one with the makings of a statesman. He saw his beloved Iceland a victim to helpless mismanagement; its inhabitants poor, spiritless, and lethargic; and the island almost entirely out of touch with European civilization. During the second half of the eighteenth century there was probably no more miserable people on the face of the globe than the Icelanders. Life had little or no promise for them. Gaieties were to be found among the Esquimaux and the Laplanders, as with Otaheitans and Maoris, but not in Iceland. Banks's own recollections, of 1772, were that "no Icelander was seen to laugh; not that he had any particular inclination for gravity, but because nothing in the detail of his mode of living seemed ever to excite him to gaiety, much less to merriment or laughter."

The cause of this apathy, upon the unvarying testimony of observers, is found in the loss of their ancient spirit through Danish domination. The energies which trade and intercommunication with other countries might have aroused were thwarted by the laws of Denmark, which treated its colony as purely a dependency, and forbade the inhabitants to negotiate for their wants with any other than the monopolists at Copenhagen. Until 1788 a concession, at the cost of £8000 per annum, was in the hands of Copenhagen merchants ; who took annual supplies to the island in exchange for fish, tallow, skins, feathers, eider-down, etc. " The oppression of this Company in fixing their own prices, upon all they bought and on all they had to sell, was beyond what had ever before been felt ; and reduced the people to the inaction and torpid state at which we found them " (*Mem. by Banks*). The Company becoming insolvent, trade was opened to the subjects of Denmark generally, as in old times. In the last decade of the century some spirit of enterprise was reviving with the Icelanders. They were becoming once more industrious and adventurous. And it appeared to those interested in the little nation that, if liberty were quite restored to them, the Icelanders would recover the active and intelligent character of their ancestors.

When Banks was in Iceland in 1772, he found the people " universally desirous of being placed under the dominion of England. The applications made to me personally by natives of the best quality were continual. Their project was that England should wrest from Denmark the dominion of Iceland. . . . They concluded that the wealth of England could easily buy, and the poverty of Denmark would willingly sell."

Olaf Stephensen of Reikiavik was one of the natives above alluded to. His son Magnus, afterward Chief Justice of the island, was a boy of ten years in 1772 ; and there was doubtless occasional communication with Banks. On January 30, 1801, Sir Joseph wrote to Magnus what appears to modern eyes a very startling message. Whether it was prompted from the other side, or begun in the dark recesses of the English Foreign Office, is not clear. The purport of it was this :

Denmark has forfeited the friendship of England, who must now take action in detaching such of her possessions as she thinks proper. Iceland is a part of the archipelago known by the ancients as *Britannia*, and geographically should be a part of the British Empire. It is in the power of the Icelanders to simplify matters by a revolutionary blow, and unanimously hand themselves over to England ; thus avoiding the disagreeable results attending conquest.[1]

The above (draft) letter is accompanied by a Memorandum, evidently drawn up for the consideration of the King's ministers, by request or otherwise :

Sir Joseph relates his former acquaintance with the people of Iceland, and gives an outline of their commercial history. The people (he says) are ill-supplied with the necessaries of life, and would rejoice in a change of masters that promised them any portion of liberty. " The bettermost people shewed a predilection for England, and privately solicited the writer to propose to his Government to purchase the island from Denmark." They suggested he should buy a farm, as an inducement to promote the business. The conquest of Iceland would be a wise measure, and subject Denmark to the humiliation she deserves without diminishing her national resources. It would " emancipate the people from an Egyptian bondage." The population would increase under a mild government. There would be no revenue for the present. The people would have to be supplied with vegetable food. The fisheries, the new market for British productions, and the supply of seamen for the British Navy, would be the principal advantages accruing to this country.

Affairs with Denmark were at a more acute stage in the autumn of 1807. The carrying trade was at the

[1] Banks MSS., Kew.

mercy of the stronger party, and Danish ships fell into the hands of English prize-courts.

One day, in October, Sir Joseph Banks had a doleful letter from Magnus Stephensen, relating his capture on the way to Copenhagen when on board an Iceland trader ; ignorant of the new misunderstanding between England and Denmark. After being conveyed prisoner of war to Leith, he was presently allowed to go back to Copenhagen in a British warship. But the owner of the prize, together with his wife and family and servants, remained utterly destitute. This was followed by a letter from Dr. William Wright, of Edinburgh, an old friend of Banks, who implored his help. The Edinburgh folk had opened a subscription for the Icelanders. Further, there was a ship detained at Cork, and another at Yarmouth. Later on, a vessel was driven into Stornoway by contrary winds and detained. All these unlucky mariners were writing and interviewing Sir Joseph Banks during the winter. Although suffering badly from gout, he not only devoted himself to their relief, but liberally helped them from his purse.

Lord Hawkesbury to Sir Joseph Banks.

" *November* 29, 1807.

" MY DEAR SIR,—I return you Mr. Stephensen's letter. I have communicated it to the King's servants, and they are of opinion that you should have some further communication with him ; with a view of ascertaining whether through him or any other channel, the island of Iceland could be secured to His Majesty, at least during the continuance of the present war. In that case the Fisheries and the Trade of Iceland would be protected ; and I should hope that we might be able to obtain the services of some of their mariners. There will be no objection, as a measure of conciliation, to releasing the few Icelanders' ships which are at Leith."

Many weeks elapsed before anything practical was done. The legal Mind and the official Back vied with each other in obstruction, while most of the poor culprits were in London nearly destitute. At length, Sir Joseph was urgent, after receipt of a lawyer's letter of a somewhat officious character, to have the affair settled. He was still disposed to drastic measures : " I conclude it will be judged necessary to annex the island of Iceland to the British Crown without the least delay," he says to Lord Castlereagh ; at any rate, it would be a wise policy to separate entirely the Danish and Norwegian vessels from those of Iceland. (On this basis, a proclamation was issued two years later, ordering that the islands of Faroe and Iceland and the settlements on the coast of Greenland, and the inhabitants thereof, be exempt from hostile attack on the part of His Majesty's forces and subjects. Ships were bound to make a port of call at Leith or London in order to evade capture.)

On February 26 Government decided to restore the Icelandic prizes already in hand ; but this was not accomplished until about April 7. English traders were allowed to trade by " license " with Icelandic ports.

These things made a profound impression on the people of Iceland ; and the proceedings of Sir Joseph Banks in connection with them aroused their eternal gratitude. Beside this, their national life was stimulated. They never relapsed into their former apathy. They began to take more interest in their relations with the outer world. A modern Icelandic literature came into existence, and in the course of another generation they succeeded in asserting some degree of political independence.

In the summer of 1809, Mr. William Jackson Hooker made a botanical tour in Iceland. He took with him a kindly message from Banks to the aged Olaf Stephensen, together with a present of books and prints. Stephensen asked " a hundred questions about him in the most affectionate manner." Then he related anecdotes of what passed during Banks's stay in the island thirty-seven years ago, in a way " which at once convinced us of the excellence of his memory, and of his gratitude and high esteem for the benefactor of Iceland." Hooker was told how most of the islanders imprisoned during the war had been supplied with money until they could return to their own country. Wherever he went on his travels, the fact of Hooker visiting Iceland under an introduction by Sir Joseph Banks opened the hearts of all, and made his progress one of overwhelming hospitality. They could not show enough pleasure and alacrity in paying attention to him. And, " if at any time we flagged in drinking, ' Baron Banks ' was always the signal for emptying our glasses." The more learned classes in Iceland were strongly affected by the memory of 1772 ; and several poetical addresses were printed in honour of the visitors.

Mr. Hooker's sole piece of bad luck was very bad. He lost all his botanical treasures by fire on his way home.

Sir Joseph Banks to Mr. W. J. Hooker.

" REVESBY ABBEY, *October* 1, 1809.

" MY DEAR SIR,—I condole with you sincerely on the failure of an expedition, from which, however, you must derive honour in the opinion of those who know you, and no inconsiderable accession to their good opinion of your talents and your resources. The decision with which you entered into the undertaking, the promptitude with which you carried your preparation into execution, and the alacrity with which you encountered the hazard of a voyage, are deeply impressed upon my mind, and have decided me to attempt to gain your friendship which

in future I shall omit no opportunity to cultivate. I condole with you for the losses you have sustained, but
much more seriously with your friends, who must lose
the participation you had prepared for them in scenes
of natural wonder and delight. . . .

"Some one must go next year to Iceland. . . . I
had once a proposal for a cargo of Scotsmen with Sir
George Mackenzie, of Coul, at the head ; but I advised
them to wait for another season.

"Recommend me to my friend Dawson Turner when
you see him. . . ."

Sir George Stewart Mackenzie, F.R.S., paid a visit to
Iceland in the following year. His mission was minera-
logical, but he contrived to gather a few plants to replace
Hooker's losses. He published a capital account of his
voyage, painstaking in every way ; one of those books
which can never get out of date. Sir George called upon
Olaf Stephensen, at his retirement in the isle Vidoe, near
Reikiavik. He describes his visit, and hospitable reception, and the costumes and the ways of the household,
with much vivacity and picturesqueness. Writing to
Banks (May 20, 1810) he says of Olaf that he is in very
good health for a man of his age. "He speaks of you
with rapture and remembers you with affection. He is
a delightful old man, and has been uncommonly polite
to me." Magnus Stephensen was not less zealous than his
father in welcoming the visitor. "We found every one
emulous in offering his services ; and I shall ever remember with gratitude the kind attention and hospitality
I experienced during my stay in Iceland, both from the
natives and from the Danes."
Sir George Mackenzie did not relax his interest in
Iceland. We find him, in 1815, writing to Banks concerning Rask's efforts to keep alive the study of the
ancient language ; and about, which in his eyes was of

higher importance, raising the standard of intelligence
and of enterprise.

The affairs of Iceland were not easily lost to the attention of Sir Joseph Banks. An entirely new sensation
arose at the period of Hooker's visit.

There was one Jörgen Jörgensen, an Anglo-Dane, who
had served as able seaman under Flinders (1802-5), and,
according to all report, a first-rate able seaman, in the
tender-ship *Lady Nelson.* He saw a good deal of life and
adventure in New Holland and the adjacent seas ; and
doubtless the fame of Banks was familiar to his mind,
seeing the associates that he had. He brought to England, in 1806, two New Zealanders and two Otaheitans,
and introduced them to Sir Joseph, who " cheerfully
took charge of them, defrayed all their expenses, and
placed them under the care and tuition of the Rev.
Joseph Hardcastle."

Jörgensen presently returned to Copenhagen. We
find him, in the following year, captain of a Danish
privateer, prize to H.M.S. *Sappho,* after a sturdy fight.
Coming to London as a prisoner under parole, he lost
no time in paying his respects to Sir Joseph Banks.
At that time, Banks was concerned in the providing food
and stores to the Icelanders. There was a Mr. Phelps,
a London merchant, who undertook this trade ; and in
December, 1808, he made a successful venture, with
Jörgensen in command. The island was simply neglected
by the Danes, and our visitors had a " most grateful
welcome from the starving inhabitants." A second
voyage was undertaken early in 1809. Mr. Phelps himself
was on board (perhaps as captain), with Jörgensen for
supercargo. The ship also carried Mr. W. J. Hooker,
who had been induced by Banks to make the botanical
tour in the island, already mentioned.

The *Margaret and Anne,* bearing letters of marque,
and carrying a shipload of provisions and dry goods,

reached Reikiavik on June 21. The people were over-joyed, but the Danish authorities met our adventurer with a stern refusal of any trading privileges. Whereupon Jörgensen seized the powers of Government, issued a proclamation to the natives and led a small Revolution. All this with the cheerful acquiescence of the natives, who supported his pretensions ; and without a drop of blood being shed. Count Trampe, the Danish Governor, was arrested and taken on board the ship.

"All the measures I adopted in my new character of Monarch of Iceland" (Jörgensen says) "partook of the character of popular reform. I established trial by jury, and free representative government. I relieved the people of one-half of the taxes, making good the deficiency of revenue by levying a small duty on the import and export of British goods, to which I had thrown open the port. I augmented the salaries of the clergy, from the bishop down to the humblest curate. . . . I advanced money for the benefit of the public schools and the fisheries. . . . I released the people from all debts due to the Crown of Denmark. . . . I erected a fort for six guns, and hoisted the ancient and independent flag of Iceland."

A few weeks sufficed to end his affairs. H.M.S. *Talbot* arrived, and put an end to what was a benevolent and well-meant, but utterly illegal scheme. Count Trampe was released. Jörgensen was made a prisoner. And the whole merry party (for they appear to have been friendly all round) prepared for departure toward London. The *Margaret and Anne* left Reikiavik on August 25, with Mr. Hooker and his botanical collections on board. Two days afterward this vessel was found to be on fire ; and as she was laden with oil and tallow, there was little chance for anybody to escape with life. But the *Orion*, prize, with the unfortunate ex-captain on board, caught up the blazing vessel just in the nick of time. Jörgensen once more took the lead. By his activity and intrepidity all the people on board were rescued. They were eventually brought home in the *Talbot*.

Jörgensen hastened up from Liverpool to London and called upon Sir Joseph at once. But he found the captain of the *Talbot* had forestalled him, with his own version of the affair, and his own ideas of the Dane's qualifications for personal government. Three weeks later, Jörgensen was arrested, charged with having broken his parole, and sent to Tothill Fields prison. He spent a year on one of the hulks in Chatham harbour.[1]

Banks appears to have befriended the Danish adventurer as far as he could. And he was not clear in his own mind about the character he had to deal with. Jörgensen was brave and rather reckless, but well-meaning. He learned gambling on board the prison-hulks, and this weakness pursued him for many years. He was ruined over and over again by getting into the company of gamblers. Thus was thrown away once more a very fine character. He ended his years in Tasmania, having been transported for life.

Sir Joseph Banks, writing to Hooker shortly after the above-mentioned incidents, says, " Whether the usurper best deserves to be hanged is a matter that may be doubted. But that both of them deserve it richly for retarding the destinies of an innocent nation in trying to facilitate their trade is what I do not doubt, and a matter of which I think you will in due time be convinced, if you are not already."

Mr. Hooker writes from Halesworth in reply (July 27, 1810) :

". . . I must, Sir Joseph, once more crave your indulgence whilst I bring forward the name of Jörgensen. I am fully sensible that you have done for him more

---

[1] The entertaining details of the Icelandic " Revolution " are to be found in Hooker's published *Journal* (London, 1813), and in Sir George Mackenzie's *Travels in Iceland* (1810).

than could reasonably be expected ; and more, too, than he perhaps deserves. I think, too, from some hints that have escaped you, that you pity him in his present situation, and that you would not object to hear of his immediate release ; but, at the same time, that it would be neither consistent nor proper for you to do any more for him. . . . I feel my situation with regard to Jörgensen a very peculiar one. His exertions on board the *Orion* most undoubtedly saved my life and those of the rest of the passengers of the *Margaret and Anne*. Besides that his pleasant manners and goodness of heart have excited in me a friendship for him which I would be glad to make use of in his behalf, but which I will never do if contrary to your wishes."

Jörgensen reappeared in 1813, called upon Mr. Hooker at Halesworth, and was entertained there. Here (he says) in " the quiet retirement of this country residence," he shut himself up and " wrote an account of the Icelandic revolution, and presented it to Sir Joseph Banks."

Banks read the papers sent him, and appears to have thought there was much to be said on both sides. He tells Hooker that the man has opened a correspondence with him; has written a story of the Revolution, and " wishes to place it in my library. I said that if on perusal I did not find anything improper I should comply with his request." A month later he writes that he has no further concern with Jörgensen : " his good or ill-fortune will be alike indifferent to me. It is not my wish to exercise the office of an avenger, and I thank God it is in no shape my duty to interfere in his case further than I have done." [1]

---

[1] The papers handed to Banks are doubtless those preserved in the British Museum (Egerton MSS., 2067–69). Since writing the above we find that they have been partly utilized in literary form : *The Convict King, being the life and adventures of Jörgen Jörgensen, monarch of Iceland, Naval Captain, Revolutionist, British Diplomatic Agent, Author, Dramatist, Preacher, Political Prisoner, Gambler, Hospital Dispenser, Continental Traveller, Explorer, Editor, Expatriated Exile, and Colonial Constable.*—By J. F. Hogan (London, 1876).

CHAPTER XVI

THE RISE OF NEW LEARNED SOCIETIES

IT now happened, with the widening knowledge of the Natural Sciences and their growing popularity as educational influences, that specialization tended to produce new groups of men desirous of a base for work independently of the Royal Society. Hence arose several new Societies, each of them on a new footing of their own. The Chemical Society and the Geological Society were among the first, and they speedily justified their separate existence. But Sir Joseph Banks was not altogether content that Fellows of the Royal Society should transfer their energies into new and independent Associations ; and his countenance was not willingly given. His views on the matter appear to have been supported by some of his friends.

Mr. Weld alludes (*History*, II, 230) to this raising of new and literary and scientific Bodies ; and tells us that it was customary, "when any of these applied for a charter of incorporation, to send a copy of the petition to the Royal Society, in order that the Council might make any objection that they thought proper." This statement throws a little light on the situation. Once having got into the way of appealing to the Royal Society, that is, to Sir Joseph Banks, it was a facile step to submission to his authority. The point that interests us is, that the augury of a sort of autocracy arising soon after the first accession of Banks to the chair in 1778, was fulfilled. As this supreme authority in the world of Science was

maintained, not by arrogance, but by habitual courtesy
in consulting the feelings of others, and by devoting his
whole time and his fortune to the progress of knowledge,
he was enabled always to maintain a personal influence
over the philosophic world. " Every one looked up to him
as a friend and counsellor. He succeeded in keeping in
abeyance among them those feelings of jealousy from
which even those who, standing apart from mere vulgar
pursuits, devote themselves to the acquisition of knowledge, are not altogether exempt."[1]

Sir Joseph Banks was not alone in his short-sighted
view, that the Royal Society would suffer by the creation
of new associations to take over part of its work. But
it was a baseless fear. Although there was some little
slackening in its energy in the later years of Banks's
presidency, the Royal Society continued afterwards in
great vigour, as a glance at its physical, astronomical,
and mathematical papers will show, or a recollection of
the names of Babbage, Davy, Baily, Olinthus Gregory,
John Herschel, etc. Indeed, one of the consequences
of the rise of new Societies lay in the occasional collaboration that occurred between them and their respected
parent.

There was in existence, however, good proof that these
notions of rivalry were erroneous as to their bearing on the
stability of the older Society. Not only with Banks's
concurrence, but with his financial aid, the Linnean
Society was set on foot in the year 1788. It had been
successful from the first, and that with topics hitherto
dealt with by the Royal Society. In its specialization
it had attracted members largely from a class of men
who were without ambition to become Fellows of the
Royal Society. In short, it was a sound and healthy
branch.

The origin of the Linnean Society was in this wise.

[1] Sir B. C. Brodie : *Autobiography*, p. 73.

After the death of Linnæus, his entire collection of books, papers, cabinets, and herbarium was offered to Sir Joseph Banks for the sum of one thousand guineas. There was breakfasting with him, when the message arrived, James Edward Smith (a young pupil of Dr. Hope), who was recently come to London, and welcomed in the coterie of Soho Square. Smith's father was a member of the renowned Norwich circle, and was prepared to smooth the young man's path in the study of Natural Science.

Banks was not disposed to make the purchase, but strongly counselled his friend to do so. Other men urged him. Mr. Smith, senior, after some hesitation consented to help his son to conclude the bargain. A bargain it really was : two thousand volumes of books, about fourteen thousand plants, and seven thousand shells, insects, mineral specimens, and birds in glass cases. After delays and dangers, and an escape from the design of the King of Sweden to intercept it, this precious cargo reached the Thames in October, 1784.

Dr. Smith presently resolved on forming a new institution for the exclusive promotion of Natural History studies. Thus arose the Linnean Society of London. He became the first President. In two years the Society had over one hundred members and associates. Banks gave it his hearty support. With his usual liberality, his purse was opened for their assistance. For example, he bore the entire cost of the copper and engraving of the twenty plates in the first volume of the Society's *Transactions*.

In the case of the Geological Society, Banks withdrew from membership ostensibly on the ground that the Council had deviated from its first principles. Really, Sir Joseph wanted the Society to be as an assistant association to the Royal Society. The body of the members preferred entire independence. Apparently there were no breaches of friendship over the matter.

S

Ten years after this, the Astronomical Society was set on foot. This greatly disturbed Sir Joseph. He strenuously opposed it " on the ground that such an association, by robbing the Royal Society of many of its members, and affecting to engross one of its most important departments, struck a severe blow at its respectability and usefulness." [1] Several persons held aloof, including the Duke of Somerset, who was first invited to become the President of the new Society— on the ground that it was not approved by Sir Joseph Banks. These alarms were needless, and it soon appeared that the Astronomical Society had an important function to fulfil, and that it was possible for gentlemen to become active members of both Societies.

The Royal Institution was founded in 1800. It originated from a suggestion of Count Rumford, who had been of late years in London, taking part in the scientific and social activities of London. [2] When Rumford was formerly resident in England (then plain Benjamin Thompson) he introduced himself to Banks on the score of a learned

---

[1] Sir John Barrow : *Sketches of the Royal Society*, etc.

[2] Rumford was now accompanied by his only daughter. There is an interesting memorandum of hers relating to this period : " My father was often at the Royal Society, and intimate with the president, Sir Joseph Banks. I would be invited to the dinners Sir Joseph gave to the select ones of his royal learned society. Through the kindness and civility of Lady and Miss Banks, his wife and sister, I several times found myself one of their party. Lady Banks was so kind, and most likely out of civility to my father, she would allow me to be with her for days together, taking me about with her, letting me see things, trying to amuse me. I recollect she took me to a Lord Mayor's ball, where I saw the princes and Royal Family for the first time. As may be supposed, the select dinners of the Royal Society were highly interesting, and where I think ladies were seldom or never admitted. I was allowed to accompany Lady and Miss Banks as a mere nobody ; but this did not prevent my making observations which never have and never will be forgotten. The idea of very learned people suggests pedantry. At these dinners there was nothing of the kind, differing only from other refined society when remarks were made to convey, perhaps, new ideas, discoveries, or highly entertaining instruction ; sometimes there being no such talk at all " (H. Bence Jones : *The Royal Institution*. London, 1871, p. 58.) This young lady (afterwards known as Sarah Countess Rumford) presently attracted the eye and heart of Dr. Blagden. But her father would not consent to their union.

scientific essay that he brought with him. They speedily became intimate. Thompson became a F.R.S. in the following year (1779), and they had corresponded while the young man was making his fortune and fame at the Bavarian Court.

Rumford must be always active in something useful, especially in improving the economical and domestic condition of the poorer classes. He was an essentially practical man, a thorough Utilitarian, before that title was acquired by the advocates of utilitarianism who made it famous. The idea now suggested itself to him to popularize Science by lectures and by laboratory work, and to bring such of the Royal Society's activities into practical paths as were likely to be embraced by unscientific people. He proposed a new Institution with this object in view. At a meeting held at Banks's house in Soho Square, the thing was adopted; and a large number of gentlemen subscribed fifty guineas for the purpose of founding the Institution. The definite object was, " For diffusing the knowledge, and facilitating the general introduction, of useful mechanical inventions and improvements; and for teaching, by courses of philosophical lectures and experiments, the application of Science to the common purposes of life." The delicate question was raised for a moment, whether its functions would not interfere with existing Societies; but this did not acquire any weight, and was, perhaps, only tendered as a courteous recognition of the valuable work which the Society of Arts had been doing for nearly half a century, very much on the same lines.

The first President was the Earl of Winchilsea and Nottingham. Sir Joseph Banks was a Vice-President. Nine managers were chosen, three of whom would retire every year, but were eligible for re-election. Sir Joseph Banks was elected for two years, and again in 1802 for three years.

The Royal Institution had a great and an immediate success. It was fashionable. Crowds of people came, and were delighted at the mixture of amusement and instruction which was offered to them, " the female part of the audience " especially. Sir Humphry Davy gave a course of lectures at the Institution in 1801.[1] His phenomenal success as a chemist and physicist, and his personal popularity, carried the Institution along with him; and, it may be said, the Institution gave Davy his opportunity. Chemical research and electrical studies now became matters of popular interest. And it looked as if the Royal Institution supplied an indubitable want. It was designated " the workshop of the Royal Society," and, as Mr. Weld says (II, 234), it deserved the appellation.

The Royal Horticultural Society was the idea of John Wedgwood, of Betley, in Staffordshire, a gentleman devoted to horticulture and a good naturalist, a friend of T. A. Knight. They were dissatisfied with the prevailing habit of leaving all to the gardener, who generally pursued the dull routine of his predecessor, without science and with little intelligence. They felt sure of being able to improve almost every esculent plant or fruit by the adoption of system and foresight in garden operations. Wedgwood suggested a meeting for gentlemen holding similar views, in order to start a Society for the study of Horticulture. Accordingly, there met at Mr. Hatchard's shop, on March 7, 1804, Charles Greville, Sir Joseph Banks, Messrs. Salisbury, Aiton, Forsyth, and Dickson, with John Wedgwood. Thus was founded, with the assistance of a number of gentlemen, the most distinguished for their love of horticulture, a Society destined to revolutionize the art of gardening, and to popularize it among all classes of Society.

---

[1] Davy's first start in life was with Dr. Beddoes, at the Pneumatic Institute near Bristol above mentioned.

The Earl of Dartmouth was the first President. Banks was one of the Vice-Presidents. Wedgwood acted as Treasurer. In 1809 the Society obtained a Charter of Incorporation. From the very beginning it justified its existence; and it exists to-day, one of the most splendid legacies of that awakening period.

At a time when Banks was loosening his hold upon some of the threads that had been so long in his hands, the Horticultural Society came as a fresh spur to activity. His love of experiment had new scope, in association with others, for the trial of ingenious improvements in gardening. His house at Spring Grove, Isleworth, had new attractions in the yearly increase and novelty of its garden. Perhaps he did not have it all his own way ; for he says, somewhere, " my gardener, who of course is my master."

Sir Joseph was heart and soul with the new Society. He read several papers at the evening meetings, all of the most practical character. A very good one was "On Inuring Tender Plants to our Climate," when he was enabled to prove, from his own experience, that exotics, however weakly and unpromising at first, mostly increased in vigour every year if their growth was carefully studied. " Old as I am" (he said) " I certainly intend this year to commence experiments on the myrtle and the laurel." Another good paper concerned the " Revival of an Obsolete Mode of Managing Strawberries." This was none other than the practice of mulching the plants with straw, a hitherto forgotten mode which he had found in an old herbal. A third paper worth mentioning told of his great success in cultivating the American cranberry at Spring Grove.

Thomas Andrew Knight was an indefatigable member of the Society. He succeeded Lord Dartmouth as President, and held the chair for a long period of years. He was long a close friend of Banks. His letters on gardening bulk largely in the Banks *Correspondence* preserved at South Kensington.

There was once a proposal for a Belles-Lettres Society. It could have been only from the prevailing habit of approaching Sir Joseph Banks on any great project whatever that tended to mental culture, that induced his friends to ask him to join on such an occasion. Mr. H. F. Greville, writing from Brighton (June, 1807), puts the question to him, on behalf of himself, Richard Payne Knight, Lord Abercorn, and other men of taste, who are concerned in promoting the scheme, and incidentally raises the question whether it would be proper to hold the meetings of the Society under the same roof as a place of public entertainment.

Banks's reply is curious as a piece of self-revelation, very rarely displayed in his correspondence with his friends. The first-person-singular seldom occurs in his letters, except as delivering an opinion or expressing his wishes. We have had to judge him from his deeds, and not from his confessions. For once, we get a faint glimpse of the man himself from his own words.

Sir Joseph Banks to Henry Fulke Greville.

" SIR,—I am thankful to you for your letter of the sixth instant, which I should have answered sooner had the printed paper which came to my house some days ago been at hand. From the tenour of the printed paper I conceived that I was invited to become a Member of an intended Society, and that the Room in which the Society was to meet was to be occasionally employed for other purposes. Your letter has shown me my mistake. I find from it that new rooms are to be built and exclusively reserved for the use of the intended Society, and that the Members of it are to be chosen by a Committee.

" I must beg on any terms to exculpate myself from all idea of my having ever considered the amusements of the great world as frivolous. I respect, I assure you, Sir, the recreations of the public, and look up to those who direct them with taste and with judgment. I see no objection to a room being occupied on one day by a Society of Philosophers, and on the next day a company of masqueraders ; and the less so as it will probably on both occasions be frequented by the same persons. I was myself an attendant on plays, operas, concerts, masquerades, etc., till prevented by infirmities ; and was my health now restored to me I should again be partaker, in the decline of life, of those gaieties which added so much pleasure to the commencement of it.

" I am sorry I cannot accept your obliging invitation to dine on the 28th, as I shall then be in the country. But this is of no moment, and I know myself too well to suppose myself a proper member of a Society for Belles Lettres. I am scarce able to write my own language with correctness, and never presumed to attempt elegant composition, either in verse or in prose, in that or in any other tongue. It is fitting, therefore, that I continue to confine myself, as I have hitherto done, to the dry pursuits of Natural History, etc."

In the case which may now be mentioned, there is still another trace of Banks's unwillingness to venture on matters to which he was wholly unaccustomed. The reasons given for his decision are sound enough, but one cannot doubt the existence of a sense that he was unequal to the post of a Judicial Inquirer. Two gentlemen wanted the assistance of some person who would act as Arbitrator, and sent word to Banks asking him to undertake it. The reply is what might be expected from one who was beginning to feel the excessive burdens of life.

*Sir Joseph Banks to Sir William Pulteney
and Sir Mark Wood.*

"Soho Square, *June* 25, 1802.

" Gentlemen,—I beg you to accept my best thanks for the flattering manner in which you have been pleased to place your confidence in me, and for the precautions you have taken to prevent unnecessary waste of my time in the investigation of the business you wish me to undertake. I must, nevertheless, trust that you will excuse me, if I decline to accept the Arbitration you propose to me.

" First, because the hazard of being a sole Referee is greater in my opinion than any man ought to subject himself to voluntarily.

" Secondly, because in my opinion, matters of Right ought in all cases to be settled by the usual Tribunals of the Country ; and matters of account, which alone are proper subjects for arbitration, ought to be submitted to those persons only who are professional Accountants.

" Thirdly, because my time is at present so fully occupied, especially since I had the honour to be a member of the Privy Council and appointed to most, if not all, the active Committees of that Body, that I have more business than I can properly attend to ; and cannot therefore undertake anything in addition without a certain sacrifice of some part of what I have already taken upon myself to perform.

" Either of these reasons will, I hope, justify me for declining an office which, in the first contemplation of it, appears almost like a moral duty. But if to this I add that experience has too frequently shewn the infinite improbability of an Arbitrator of the most enlightened abilities having it in his power to satisfy both parties ; and the too frequent occurrence of both parties being equally dissatisfied, you will acknowledge that, holding, as I hope to do, a place in the esteem of two such valuable men as you are, I am justified in declining to put to the hazard what I have so much reason to value."

It was, perhaps, a surprise to Banks himself to be consulted by the Prince of Wales on some manuscripts from Herculaneum which had been sent to him, and whose first thought was that the President of the Royal Society should undertake the examination of them.   But Banks did not shrink from the task.   He forthwith superintended the process of unrolling and deciphering the documents.   Sir Thomas Tyrwhitt (December 23, 1804) writes to assure Banks of his thanks for his "zealous assistance, without which progress would have been much slower."   This also was a very protracted business. It included a chemical examination of the manuscripts by Humphry Davy.   Not until February, 1819, was a Report officially made.

CHAPTER XVII

REVIVAL OF BOTANICAL EXPLORATION

THE times were unpropitious for the dispatch of Collectors on behalf of Kew Gardens. An opportunity came, however, in the year 1803, which resulted in developing a new import of exotic plants from China. This was continued with success for several years, accompanied by the reverse operation of transferring English fruits, etc., to China.

The botanist was William Kerr, son of a nurseryman at Hawick. Mr. David Lance, an official of the East India Company's factory at Canton, was an intimate friend of Banks ; and the occasion of his departure was taken for Kerr to go under his protection. Kerr proved a valuable aid to the cause of horticulture, and a good correspondent for Banks, who treated him with the constant courtesy he extended to every one in his employ, especially gardeners who were expatriated in the cause of Science. Their correspondence is rather technical for these pages. But one enclosure of Kerr's is of remarkable interest, as showing once more how the votaries of Science could unconsciously serve the cause of international friendship :

Puankhequa (President of the Company of Merchants privileged to trade with foreign merchants at Canton in China) to Sir Joseph Banks.

" *10th day of the 1st moon of the 11th year of Kia King, or*
" *the 28th February,* 1806.

" SIR,—The celebrity of your name has been long known to me, as by Mr. Lance I have been informed of the respect due to your distinguished merit and abilities ; but the letter and presents with which you have lately honoured me I particularly esteem as a prelude to a nearer and more intimate acquaintance with you.

" It is extremely gratifying to me to find that my endeavours to assist Mr. Lance, and his Britannic Majesty's gardener, in the highly useful and interesting pursuits in which they have been engaged, have proved acceptable. But I blush to receive for so trifling a favour the very elegant return which you have made me. I nevertheless readily accept of those presents as a testimony of your esteem and regard. My apartments will be adorned and my table will be graced by the several articles, and they shall be so disposed as may appear most worthy and honourable to the magnificent donor as well as recall him oftenest to my remembrance.

" If my country affords any natural or artificial productions which may be curious or interesting in your eyes, I trust you will inform me and signify your commands ; for in endeavouring to execute them I shall have a peculiar pleasure.

" In the meanwhile I send a few presents which, being more remarkable for their rarity and curiosity than for their value or magnitude, I trust you will not hesitate to accept, as a mark of the esteem and consideration with which I have the honour to subscribe myself," etc. etc.

" List of articles :

" No. 1. A pair of large rosewood and glass lanterns, ornamented with silk Tassels.

" No. 2. A pair of large horn lanterns, ornamented with silk Tassels.

" No. 3. A set of 20 cups and covers of new and curious Porcelaine.

" No. 4. A set of 20 enamelled ornamental stands for ditto.

" No. 5.   A set of lacquered and inlaid dishes or
waiters.

" No. 6.   Four red boxes, varnished and carved in a
rare and curious manner.

" No. 7.   A peculiarly curious and ancient dwarf Tree.

" No. 8.   Eight pots of the finest moutans."[1]

After seven years' work at Canton, interrupted by a
botanic trip to Luzon and the neighbourhood, Kerr was
recommended by Banks to the Governor of Ceylon, who
wanted to find a competent superintendent to the new
Botanic Garden there.   Kerr set it on foot, and was
making a flourishing affair of it, when he was seized with
some illness incidental to the climate, and died in 1814,
only two years after his arrival.   Alexander Moon
succeeded him, arriving out in 1817.   The site of the
garden was removed to a more suitable spot, and continues
to improve.   Favourable reports of the Peradeniya garden
are still received periodically in London.

This mission of Mr. Kerr to Canton is the more interesting because of Banks's association with others
concerned in the Canton factory.   Sir George Thomas
Staunton, son of the famous ambassador to China, in
company with Lord Macartney, was in the beginning of a
life of splendid usefulness ; at present supercargo at the
factory.   Between Banks and Lance and Staunton
there appears to have been a common regard.   Staunton
was one of the new generation upon whom the mantle of
Banks's fellows was to fall.   To such persons Sir Joseph
held out his hand of encouragement without reserve
and without parade.   He lived to see Staunton risen to
high distinction as a diplomatist and a linguist, and the
first authority on China and the Chinese.

Sir George was asked to concern himself with the welfare

---

[1] i.e. the tree-pæony.   In his covering letter, Kerr says the Chinese
gentleman had known the dwarf tree for thirty years, and it was
supposed to be at least one hundred years old.

of Mr. Kerr; and he forthwith made this his obligation. The botanist had good reason to be thankful for this patronage, in his very difficult surroundings. Much of his success was owing to it, and he gratefully acknowledges to Sir Joseph Banks the assistance and encouragement he has had.

Another young man who made afterwards something of a reputation went out to Canton in 1806. This was Thomas Manning, the friend of Charles Lamb, who had been released from detention in France through the intervention of Sir Joseph Banks. Manning's friends thought it a sort of freak; but he became interested in Chinese matters, and saw that the only way to learn the language and understand the people was to visit the country. With Banks's interests in the East India Company, he was enabled to go under agreeable circumstances, as a doctor to the factory at Canton, with free passage out, and a residence. He was entrusted with a number of European plants for the care of Mr. Kerr.[1]

[1] A long and amusing letter from Manning (posted at Cape Town, Aug. 9, 1806) is included in the Banks MSS. at Kew. One extract is worth making : " After a pleasant passage of ten or eleven weeks we are arrived within a few days' sail of the Cape. . . . I am exceedingly comfortably situated, and treated with great respect and even distinction. My greatest want is good society. I am among a set of grossly ignorant people. The rogues soon found out my superiority of acquirements, and they now will give me credit for knowing what I am really ignorant of. Because they see I take a great interest in the plants, they look up to me as a botanist, and I cannot undeceive them. The plants are most of them thriving, but some are dead, or at least have lost their leaves. . . . I have been careful to shelter them from all violence, either of heat, rain, or wind ; and the captain's steward has been particularly careful in tending them. I had a little trouble at first in preventing certain officers from plucking the odoriferous leaves, but a little gentle expostulation and management soon succeeded. The first mate does not approve of having a garden on the poop at all. He says it racks the ship all to pieces. The Captain agrees in the same story, and when the beams creak in the cuddy, they turn to me sometimes and d—— the flower-pots. But 'tis half in joke. At least, they exaggerate the matter. . . ." Living plants are but sickly passengers at sea. A consignment of boxes from Kew to Canton in March, 1805, contained living plants as follows : 15 grapes, 24 plums and cherries, 24 peaches and nectarines, 20 pears and apricots, 19 figs, etc., 12 rhododendrons and azaleas, 34 roses, 110 various bulbs, 22 pelargoniums, and 64 miscellaneous. Few of them reached their destination, on account of accident or neglect.

Sir Joseph Banks to Sir George Staunton.

"SOHO SQUARE, *May* 7, 1806.

" MY DEAR SIR GEORGE,—The bearer, Mr. Manning, has for some time back destined himself to the almost impossible task of travelling in China, in order to gain a real and substantial knowledge of all that that enigmatical nation are acquainted with ; and has long resolved that obstacles shall be surmounted, difficulties despised, and even Death itself held in contempt ; in pursuit of an enterprise more hazardous, perhaps, than any which the sons of Science have engaged in during the period of our lives.

" To you, my dear Sir, I beg leave to recommend him, as a man for whom I feel a warm friendship, which I trust nothing but real merit could inspire. He possesses much talent, much information, and much real Science, as well as a disposition eminently suited to the surmounting of difficulties. And, beside this, as many amiable dispositions of temper and conduct as any idle sons of the town with whom I happen to be acquainted.[1] . . . You are, I understand, to be visited by a missionary either this year or next. I have not been able to learn much about him, except that he belongs to the eclectics, who are an association of every kind of person that dissents from the rites of the Established Church. . . .

---

[1] Banks was rewarded for the pains he took to forward Manning's enterprise, in securing a new and interesting correspondent. Sir George Staunton showed the greatest attention to Manning, "giving him instruction and assistance in the most obliging manner " ; he was thus enabled to get in closer touch with the Chinese than was usual for an Englishman. He persevered in his attempts on the language, but amid much opposition, for the natives would not help him. Among other troubles with the Mandarins were the repeated efforts to make him cut off his long and ample beard ; " I would rather go and live in the Bonze house over the water, and see no Europeans at all, than part with my dear beard." He offered his services as Astronomer to the Emperor and Physician to the Empress, but the Mandarins withstood him. Manning's journal to Lhasa, with some notes of his life, was published by Admiral C. R. Markham (London, 1876). A very entertaining volume.

My own opinion leans to the probability of his obtaining the crown of martyrdom much more readily than a single convert to the Cross."

The new missionary went out in the following year : no other than Robert Morrison. Far from incurring martyrdom, he captured hearts everywhere, even among the higher Chinese.[1] Sir Joseph Banks introduced him to Staunton, and commended him to such friendly countenance and aid as his situation in the factory would permit.

Perhaps a closer bond between Banks and his friends at Canton was porcelain. Lady Banks was an ardent and intelligent collector. Banks himself became a by no means inadequate authority on the subject. According to Sir John Barrow, he left among his papers a " curious, interesting, and well-written history and art of the manufacture of porcelain by the Chinese, illustrated by a very select and extensive collection of choice and variegated specimens that were in the possession of Lady Banks." Doubtless this MS. is still in existence, hoarded by a Collector of another sort, and it would repay examination. For the present, some of our readers will peruse an extract from Banks's first letter to Mr. Lance with considerable interest.

Sir Joseph Banks to David Lance, Esq.

" SOHO SQUARE, *August* 30, 1803.

" MY DEAR SIR,— . . . At Lady Banks's desire I enclose to you the paper you will receive with this. She is a little old-china mad. But she wishes to mix as much reason

---

[1] Translator to the East India Company. Interpreter to Lord Amherst's Mission (1817). Wrote a *Chinese Grammar*, compiled a *Dictionary*, and translated the Bible into Chinese. Died at Canton, August, 1834.

with her madness as possible. She has heard much of old china in England, but does not believe that any of it is older than Queen Elizabeth's reign, and that very little indeed is old.

" She thinks that all dishes and plates made after the models of silver plate, as indeed is the case with the greater number, must be very modern ; that is, since the English traded with Canton in 1680, when I believe the first direct ship sailed from London. She has an idea also that tea-pots, and all the tea-service, are unknown to the economy of the Chinese. Coffee-pots she is sure are so. She believes, however, that the Chinese use small cups, not very unlike tea-cups, for their usual food ; and possibly tea and coffee cups for drinking *som shee*, their ardent spirit. She wishes much the same information on the subject of burnt-in china, which is said to be manufactured in the interior, and painted there with the colour of blue ; but that it is painted with red in the vicinity of Canton, and burned a second time.

" On the subject of a certain kind of china ware much admired in Europe, called here green enamel, she also is desirous of information. It is now to be found in all collections of old china. But in those made at the beginning of the century, no such china is to be found ; as, for instance, Queen Caroline's cabinet at Windsor, where nine-tenths of the china is blue and white, or white. Lady Banks knows that the old Nankin blue-and-white is in point of material much superior to all other china. She wishes to obtain, if possible, some account of the comparative degrees of imperfection in those that are ornamented with a variety of colours, as she is inclined to think that they are coarse in the bisquet in proportion as they are gaudy in the painting."

Banks's letters to Sir George Staunton usually contain some reminder on the subject ; as, " Lady Banks, who came into the room while I was writing, requests me to put you in mind of her Dairy. If any piece of odd, unusual, old, or middle-aged china should fall in your way, she will be thankful in the extreme for your remembrance." In point of fact, the pair of them had got the craze very badly. Sir George Staunton did not fail to attend to their wishes. Several consignments of valuable porcelain were sent to London, to the delight of Lady Banks and the surprise of Sir Joseph, to whom it was a matter of a new Discovery. He had no idea, until he saw these things, that porcelain was almost a lost art since European colouring and imitations came into vogue.

The return of peace, after the long contest in which nearly every part of the world was involved, was hailed by few with such welcome as by the naturalists. There would be no more dangers of capture at sea, and no longer any interference, under a "little brief authority," with the local work of Explorers and Collectors.

Some persons were fortunate enough to escape annoyance throughout the whole period of the war, as in the case of those under the protection of the East India Company. At Calcutta, Canton, Ceylon, etc., the botanists were able to pursue their tranquil career, which was, as a rule, devoted to economic cultivation for the use of the natives. But their consignments to England were much diminished, while better times were awaited. Again, a man like Sir Thomas Stamford Raffles, holding the Dutch possession of Java, was by his great abilities not only able to govern the country better than it had been governed before, and to introduce some civilization into the interior of the Island, but to investigate its natural productions. In the botanical department he was assisted by a young American surgeon, Thomas Horsfield, who had been in Java since 1804 studying the flora of the island.[1] When Raffles was appointed to the com-

---

[1] Dr. Horsfield's collections are now with the Linnean Society.

T

mand in 1811, he soon found out Dr. Horsfield, and gave the whole weight of his authority to protecting and assisting him. He was himself a botanist ; and the two together did a prodigious amount of work, the results finding their way in the end to the British Museum collection. After five years at Java, Raffles went in 1818 to command in Sumatra, and worked there a similar revolution in the order of things. Although a young man, and the envy of some of his fellow-officers, Raffles early secured the notice of Sir Joseph Banks, who naturally detected his great promise ; and their correspondence became very friendly.

At Calcutta, a new man was coming to the fore. This was Dr. Nathaniel Wallich, a Dane, who entered the East India Company's service. He was made superintendent of the Calcutta Botanic Garden in 1815, and carried it on in great perfection for many years. His letters to Banks are exclusively botanical, but they bear unmistakable signs of devotion to his pursuit.

Soon after the conclusion of peace, in 1814, the Curator at Kew Gardens made it his business to revive the question of sending out plant-collectors. He told Sir Joseph that he had in view several young gardeners, men of sound principles and invaluable zeal for the service, and having the requisite knowledge ; he wished to lay the matter before the Prince Regent if Sir Joseph agreed that the time was auspicious. Banks wrote in reply, full of his old alacrity. He thought the Cape of Good Hope and New South Wales were the most productive places for a new research, and there were other places he should wish to be visited.

Banks had now a new race of officials to deal with. Men had arisen who knew not Sir Joseph, in the sense of being an applicant for public money. There is a very good letter to Mr. George Harrison (dated September,

1814), in which he presents an elementary view of Kew affairs, for the instruction of the new people in office.

After proclaiming the importance of the Kew Gardens, and its superiority over all similar institutions, in the eyes both of Englishmen and foreigners, he protests that its position is somewhat impaired by the interruption of communications during the late war. It is desirable to renew its activity in the search for exotic plants, and, of course, it is partly a matter of public expense. " The plan of collecting for Kew, as established by His Majesty, has hitherto been to employ those young gardeners, educated in the Gardens, who showed the most inclination to, and made the greatest proficiency in botanical pursuits, and were best skilled in the successful arrangement of the plants in the Gardens. Among the many young men who work there in the hope of being recommended to gentlemen's families as gardeners when they have learned the art, some were always to be found whose dispositions led them to the study of Botany, and whose talents enabled them to excel in it. From these the best were selected ; and it is remarkable that I do not recollect one instance of a man well acquainted with the plants in the Gardens who did not feel an ambition to be employed as a Collector.

"Although the pay of that employment was regulated more by that attention to strict economy, which all who have a concern with the Privy Purse of the Sovereign are bound to exercise, than to that well-regulated liberality which those who are paid out of the Public Purse experience : who, if their salaries do not enable them to save a sufficient provision for their old age, are provided with pensions of retreat sufficient to render the latter and less active part of their lives comfortable and happy.

" The establishment of a Kew Collector was, forty years ago, £100 a year as wages. He was allowed to draw bills for travelling expenses and board wages to the amount of

£200 more ; but this he was never allowed to exceed, and, in fact, it was in almost all cases enough, and was never exceeded without a satisfactory explanation being given.   In no instance, as far as I can recall, has any censure been passed on any of those Collectors.   So well does the serious mind of a Scotch education fit Scotsmen to the habits of industry, attention, and frugality, that they rarely abandon them at any time of life, and I may say never while they are young."   Sir Joseph then proposes to give them £180, since money is much higher in value than it was.   Out of this they may be expected to save £150 at the least, as £30 it is presumed will be sufficient to furnish them with clothes and pocket-money. A free passage in one of the King's ships is expected, and mess with the warrant-officers.   Governors in the colonies are instructed to supply them with bullock-wagons, etc., from the public stores. . . . " I am willing, if Lord Liverpool thinks it would be advantageous to the undertaking, to audit the accounts sent home by the Collectors, and to certify them for the Treasury."   He proposes Allan Cunningham and James Bowie, who are ready, and even anxious, to obtain the employment as Collectors.   As though warned by the adventures of Mr. Caley, he says that the Collectors " must be directed by their instructions not to take upon themselves the character of gentlemen, but to establish themselves in point of board and lodging as servants ought to do."

After alluding to the commerce of Exotic plants, already of some importance and promising to improve, he says, " The Domestic trade in plants, supported far above its natural level by the use of growing plants in all expensive entertainments, maintains a race of sober, healthy, and industrious people, daily on the increase. This will also be increased by the introduction of beautiful novelties.   These considerations it is hoped will induce His Majesty's Ministers to foster an establishment which does honour to the Science of the Country, promotes in some degree its commerce, and aids its population. . . ."

The two young Scotchmen were sent off in October, 1814, first to the Brazils, with very satisfactory results. In 1816 Cunningham[1] went to New South Wales and New Zealand. He joined John Oxley's Expedition to the Macquarie River, which turned out extremely interesting in a botanical point of view. He remained a Kew Collector until 1831, and his work is commemorated in a grand monograph of the New Zealand flora. James Bowie went, in 1817, to the Cape, and was occupied there for several years with very great success. His plants are in the Herbarium of the Natural History Museum. Some drawings of his are in the Kew Collection.

Another emissary was Dr. Clarke Abel, who held a medical appointment under the East India Company. When Lord Amherst's embassy to China was in preparation, Banks proposed to the Company that Abel should go with him as naturalist. An ample outfit was provided for him, in apparatus, etc., and James Hooper, a Kew gardener, appointed as his assistant. This man turned out a most valuable help. His industry as a Collector was unremitting. Unfortunately he suffered shipwreck, with three hundred packages of seeds in his custody,

---

[1] Allan Cunningham is one of the most remarkable men who entered the ranks of Botanical Collectors. Born at Wimbledon, of a Renfrewshire father, he was first sent to a conveyancer's office in London ; but a fondness for botanical pursuits led him to the notice of W. T. Aiton. He became associated in the production of the second edition of *Hortus Kewensis*, and being introduced to Robert Brown, who speedily detected his talent, Cunningham remained an employee at Kew, adding to his knowledge and botanical qualifications. His life's opportunity came when Banks required a new emissary as above mentioned. Cunningham's adventures are of the most abounding interest, both in Brazil and New South Wales. His manuscript journals, and some letters to Sir Joseph Banks, are still preserved among the treasures of the Natural History Museum, together with the collection of plants made in Brazil with James Bowie ; and his own Australian plants (1818-26). *v.* also *Narrative of the Survey of Australia* (1818-22), by Phillip Parker King.

which were all lost.   Hooper afterwards became " *Hortu-lanus* " at the Buitenzorg Botanic Garden, near Batavia.

Another exploring scheme was to take the river Congo from its estuary upward in order to prove its connection with the Niger.   In spite of some objectors it was generally believed that the two rivers were one and the same. One of these objectors was Reichard, a German geographer, whose idea was that the Niger poured its waters into the Gulf of Benin.   He was right, as it turned out ; but at the period of which we are writing, Reichard's hypothesis was " entitled to very little esteem."

Sir Joseph Banks learned of this project through a letter of Mr. Barrow, Secretary to the Admiralty, dated July 29, 1815.   It was felt (he said) that an expedition up the Congo was more likely to be successful than one down the Niger.   " As to whether the two streams are one and the same, it would be discreditable for this country to remain much longer in ignorance."   In the meantime, they could do nothing without Banks's assistance ; and would he be good enough to cast about for a proper Naturalist ?   He had found an excellent fellow for commander, in Lieutenant John Kingston Tuckey, R.N., who had already distinguished himself in the seas surveyed by Flinders, and whose recently published *Maritime Geography and Statistics* showed his liking for this sort of work.

Banks suggested a steamboat for the purpose of the river voyage.   This was at first approved, but after conferring with Mr. Rennie, and Messrs. Boulton and Watt, Barrow reported that they considered it quite impracticable to build one suitable for the river.   But Sir Joseph had been aroused by the idea, and put forth all his old mental energy.   He wrote a long and reminiscent letter, deducing from his own experience the best means for making the thing a success.   He persisted in favour of a small steamboat : " Is it conceivable " (he wrote) " that a rich and powerful nation, undertaking to explore the swiftest river in the world by ascending the stream, and having newly discovered a method of stemming the currents of rivers, should neglect to use this discovery as a public measure, when individuals find it profitable to avail themselves of it in their private concerns ? "

Banks was obviously bent on assisting the Expedition, and his letter showed so much evidence of his zeal, at the same time submitting his judgment courteously to the opinions of others, that the authorities gave way on the matter of the steamboat. However, after the little vessel (named the *Congo*) had been constructed, and an engine built into her, it was found that about four knots only could be made. As it was late in the season for making a new engine, this one was taken out ; and the *Congo*, when properly rigged, turned out a very excellent sailer.

Tuckey presently departed with high expectations of being one of the world's great explorers. He took with him Lieutenant John Hawkey, who had been a fellow-prisoner in France ; also Christian Smith, a Norwegian botanist, and Mr. Cranch, naturalist. They succeeded in making two hundred and eighty miles up the Congo. Then disaster came swiftly upon them. Smith died of fever, September, 1816. Hawkey and Tuckey soon followed him to the grave.[1]

---

[1] *v. Narrative of an Expedition to explore the river Zaire, usually called the Congo, in South Africa* (London, 1818).

CHAPTER XVIII

FAILING HEALTH, BUT UNFLAGGING ZEAL

FROM an early period in the nineteenth century Sir Joseph Banks seems to have been, more or less, a confirmed invalid. An attack of gout was to be apprehended every winter. In some cases it would last over several months. He would write to Bligh, or Flinders, or other distant friends, with pathetic allusions to his growing infirmities. To those about him it was a matter of grave concern at each repeated occasion of taking to his bed.

Yet, it never seemed possible for Banks to lose sight of the numerous interests which life had for him. At the time when he was absorbed with the new threat upon his Presidency, at the close of 1802, his life was passed amid much bodily suffering. Only his intimates knew that he had " a racking cough, with aches and cramps all over," and consequent sleeplessness. During the next winter, he was in bed with gout for many weeks. His recovery from this—for which he gives grateful acknowledgment and a handsome fee to his friend Home, was followed by a long period of relief. But four years later, when he had the troubles of the Icelanders on his hands, he was again seriously ill. With all this, Banks's mind never seemed to flag. There were as many activities as ever for him ; and he was buoyant, and cheerful, and interested in life, when actually out of pain.

It must be said, however, that there was sometimes visible a deterioration in his temper, and in his manner of contradiction. Yet his judgment was usually sound and fair, to the very last.

The same round of festive gatherings and *conversazioni* went on with undiminished brilliancy. He was a constant attendant at the Royal Society Club. Considering his bodily infirmities this was very wonderful as a triumph of mind over matter.

" For fourteen or fifteen years previous to his death he lost the use of his lower limbs so completely as to oblige him to be carried or wheeled, as the case might require, by his servants in a chair. In this way he was conveyed to the more dignified Chair of the Royal Society, and also to the Club, and conducted the business with so much spirit and dignity that a stranger would not have supposed he was often suffering at the time, nor even have observed an infirmity, which never disturbed his uniform cheerfulness." [1]

Yet, in spite of his being thus crippled, he managed to live a full life, attending to his various official duties with his old zest. He took his physical troubles with more than philosophy. He rarely alluded to his growing weakness ; and then without murmuring. " I think I shall not last much longer " (he writes to Bligh). " I thank God I have had a long and happy life, and think I am quite willing to resign it. At this moment I have no use of my left hand, and not much of my legs." There were still twelve years of his life to run. And still he could write jocular letters to Everard Home, on colchicum, quackery, and *eau médicinale d' Husson ;* perhaps with a view of cheering up his fellow-sufferer. He could be merry enough, when not in actual pain ; and he kept up those social pleasures that were in reach. He writes to Home (August 24, 1818) :

" The Club prospers. Till last Thursday we had not

---

[1] Sir John Barrow : *Sketches of the Royal Society*, etc.

mustered less than nine. On the last club-day we were
seven, agreed to reduce the number from twelve to seven.
. . . Murdoch never misses. Barrow is a good attendant.
Of my attendance you are quite sure, as I have no other
dinner on Thursday when in London."

The very last years of Banks's life showed many
evidences of his vigour of mind. He was following atten-
tively the doings of his younger botanical friends : W. J.
Hooker at home, Caley at St. Vincent's, Dr. Wallich at
Calcutta, and John Reeves [1] at Canton. This last-named
gentleman was Inspector of Tea for the East India Company, and had been a faithful correspondent since the
days of William Kerr. Sir Thomas Stamford Raffles wrote
encouragingly of his own fine management of affairs in
Sumatra, and his discovery of a new botanist in Dr.
Thomas Horsfield, of Philadelphia. Captain Scoresby,
another man rising into fame, had much communication
with Sir Joseph on his projects of Polar discovery. Mr.
T. A. Knight, with his gardening experiments, and Major
Rennell, active as ever in geography and topography,
were among his older friends who kept closely in touch
with him. Another devoted correspondent was Lady
Hester Stanhope ; a good letter writer, and an excellent
narrator, who entertained Banks for several years with
gossip, and her own philosophy, from Syria.

Nor did the circumstance of Banks's physical weakness

---

[1] Reeves is a man not to be forgotten. " During the whole period of
his residence in China, 1812–31, he contributed largely to English
horticulture . . . not only by his own direct shipments but also by
collecting plants during the spring and summer, establishing them well
in pots, previous to the shipping season, and then commending them to
the care of the captains of the Company's ships, to whom he was also
always able to recommend the most desirable plants for transportation
to England, and to whom he succeeded in communicating the enthusiasm which animated himself. . . . He was either the immediate or
indirect source from which we derived the Chinese Azaleas, Camellias,
Moutans, Chrysanthemums, Roses, and numberless other treasures,
which have been for so many years the glory of English collections."—
Bretschneider : *History of European Botanical Discoveries in China*
(London, 1898).

being so notorious deter people from regarding him as the first and last authority in the scientific world. All were so habituated to seek his patronage, or assistance, or advice, that they could not forbear going to " the fountain head," as they had done for thirty years or more. And this is true, not only of the numberless obscure, with their gleams of fortune if only Sir Joseph could be induced to give them a lift. It was just the same with those of his own rank, of his own fellow-workers in science. His influence had become solid and overwhelming. There was, for one example, an attempt to form a garden in the new Regent's Park, for Botanical and Horticultural study. Dr. John Latham, a distinguished physician and botanist ; John Galt ; and some other well-known gentlemen were in the project. It actually fell through because Banks's countenance was considered indispensable, and this was not forthcoming. And Regent's Park waited for its Garden until 1839, when it was begun under the auspices of the Royal Botanic Society.

The episode of Mr. Salt and the Egyptian antiquities is one of the latest semi-public affairs in which Sir Joseph was concerned.

The *dramatis personæ* are Lord Valentia, a renowned traveller of the time ; Henry Salt, his secretary and draughtsman ; Nathaniel Pearce, their servant ; and Mr. W. Hamilton, orientalist, and a prominent member of the cultured world. Valentia and Banks were on terms of intimacy, so that an acquaintance with Salt followed in due course. Salt made another important journey, on embassy from the British Government to the King of Abyssinia ; and six years afterward was made Consul-General for Egypt.

Sir Joseph Banks, who was now giving up to Archæ-ology a full share of the energies he had hitherto lent to Natural History, was now an assiduous student of relics, antique gems, " tombstones," and such-like. He suggested to Salt that he might benefit by the facilities afforded in his situation at Cairo, to collect Egyptian antiquities for the British Museum. It was a congenial idea. Salt entertained it with some warmth ; and, when he landed in Egypt, took measures for carrying out plans of exploration, and for buying specimens and examples. He was in fairly good financial circumstances ; but zeal carried him forward, and some of his purchases were made with borrowed money. He sent an agent to Thebes to buy antiquities. He employed Belzoni to bring away the colossal head of Memnon : an object long familiar to us in the public hall of the Egyptian Department at Bloomsbury. In this matter Burckhardt shared the expenses, and the two made a present of it to the British Museum, along with other valuable items. Pearce brought from Abyssinia some trifling additions to the spoil.[1]

When Salt found that he had been at a cost greatly exceeding his anticipations, he naturally looked for some recompense. He had embarked all his little patrimony in the determination to secure a number of valuable antiquities. Hence arose the question of prices to be paid. He sent Hamilton a list of items, with what he considered a fair price marked against each. Some of these were fine specimens of antiquity ; notably among them an alabaster sarcophagus, which he valued at a very high figure.

The prices were simply his own suggestion, which it was within his right to make. But, contrary to Salt's intention, Hamilton showed the list to Sir Joseph Banks.

---

[1] Pearce stayed in Cairo until his death in 1820. He led a life of remarkably varied adventure. Apprenticed to a carpenter, he ran away to sea ; was again apprenticed to a leather-seller, and presently enlisted in the Marines. He was made prisoner by the French, served again in the Navy, found his way on board Lord Valentia's ship ; joined Henry Salt ; and at last became a botanist and curiosity-hunter. Pearce stated very positively that Mungo Park was still alive in 1818, and was in honourable captivity at Timbuctoo.

Other gentlemen saw it. Banks gave a rash and ready opinion on the matter and stuck to it angrily and obstinately. Meanwhile, the unfortunate list passed from hand to hand ; and the general opinion on Salt's conduct became unduly weighted by that of Sir Joseph, as though it were an attempt at extortion.

When the news of all this came to Salt's knowledge, he wrote from Cairo (May, 1819) with surprise and alarm. He protested to Mr. Hamilton that he had been completely misunderstood. He had offered the collection to the Government, at their own valuation ; but he had enclosed a list with his private estimate of the value of the different articles, without, however, setting up a standard for the Government to act upon. " To prevent any future misunderstanding I now take the liberty of offering, through you, my whole collection to the British Museum (except a few articles intended for Earl Mountnorris) without any condition whatever, and shall feel a great pride in hereafter rendering it complete. The expenses incurred in forming this collection have been considerable, and have somewhat seriously trenched on my small private property. Should the Trustees be pleased to reimburse me, in whole or in part, I shall receive it as an obligation ; otherwise, shall rest perfectly satisfied in the idea that my services in this respect will not be ultimately overlooked by the Government. . . ."

He stated also that Belzoni had a substantial claim to a share in any pecuniary results ; and that expert had put a price on the sarcophagus considerably over £2000. The Secretary at War, Charles Yorke, was also appealed to, as one officially interested.

Salt's explanatory letters seem to have given satisfaction to everybody but Sir Joseph ; who still adhered to his first thought, that there had been an attempt at imposition. He was not disposed to accept Mr. Salt's reasons for his having sent such a list, and long months elapsed before he yielded to the now prevailing opinion : that the antiquities offered were really of high value.

It was during the progress of this controversy that Hamilton, in a letter to Lord Valentia (now Mountnorris), used the expression that has sometimes been quoted about Banks : " Sir Joseph, you know, is a man of a word and a blow." Doubtless, he had acted hastily, and without having obtained an expert opinion. Banks could not be called an expert in Egyptian sculptures, as such ; although he was well versed in some lines of archæology. Since, however, his prestige was so great in everything that concerned the Museum, even Hamilton (who knew everything about this class of antiquities) was compelled to wait ; while he did his best to soften matters for Salt. The great probability is that Sir Joseph's powers were failing, certainly as to memory. By November, he was writing to Lord Mountnorris in a tone which simply denotes a concern that Salt's treasures be secured. "One of Salt's statues " (he says) " has arrived. It is the one he values at £800,[1] consequently in his opinion the best he has. I am taking measures to have it removed to the British Museum and placed there in public view, preparatory to the arrival of the rest. The other is now in a cellar in the city. I went with Combe in the hopes of seeing it, but found the stairs too narrow. He saw it, and considers it far the best Egyptian work he has seen, Memnon excepted. The General Meeting of the Trustees will very soon take place. I shall then make Salt's proposal, which will I hope be instantly accepted."

Banks's death occurred seven months later. The business lagged, and it was not until May, 1822, that the Trustees decided to take the Collection, according to a valuation which was forthwith to be made ; excepting the alabaster tomb, on account of the high price put upon

---

[1] " Life-like sitting figure in marble, from Carnac, a perfect imitation of nature ; the seat covered with hieroglyphics."—H. Salt.

Landing the Treasures, or Results of the Polar-Expedition.!!

G. Cruikshank, after a design by Capt. Marryat

LANDING THE TREASURES

it by Belzoni. This last precious item was brought by Sir John Soane. It remains one of the most treasured objects at his house in Lincoln's Inn Fields.

Sir Benjamin Brodie, President of the Royal Society in 1858, was another of the younger men associated with the veteran Banks. He was a pupil of Sir Everard Home, through whose friendship he was introduced at Soho Square; and presently attracted Banks's notice as a diligent student in chemistry and anatomy. He became a F.R.S. in 1810. The Copley medal was awarded him in the following year.

In his autobiography, Brodie offers some personal recollections of Sir Joseph Banks: " He invited me to the meetings which were held in his library on the Sunday evenings which intervened between the meetings of the Royal Society. These were of a very different kind from those larger assemblies which were held three or four times in the season by the Duke of Sussex, the Marquis of Northampton, and Lord Rosse, and they were much more useful. There was no crowding together of noblemen and philosophers and would-be philosophers, nor any kind of magnificent display. The visitors consisted of those who were already distinguished by their scientific reputation; of some younger men who, like myself, were following these greater persons at a humble distance; of a few individuals of high station who, though not working men themselves, were regarded by Sir Joseph as patrons of Science; of such foreigners of distinction as during the war were to be found in London; and of very few besides. Everything was conducted in the plainest manner. Tea was handed round to the company, and there were no other refreshments. But here were to be seen the elder Herschel, Wilson the Sanscrit scholar, Marsden, Major Rennell, Henry Cavendish, Home, Barrow, Blagden, Abernethy, Carlisle, and others. . . .

His London residence was in Soho Square, there being extensive premises behind his dwelling-house, which contained his library and his botanical collections. The former consisted chiefly of books on Natural History, and the Transactions of learned Societies, and was probably in these departments unrivalled in the world, His principal librarian was a Swede, Jonas Dryander. and under his superintendence the library was so well-managed that, although books were lent to men of science in the most liberal manner, I believe that not a volume was ever lost. Dryander was indeed a pattern as a librarian. The library over which he presided was to him all in all. Without being a man of science himself, he knew every book and the contents of every book in it. If any one enquired of him where he might look for information on any particular subject, he would go first to one shelf, then to another, and return with a bundle of books under his arm containing the information which was desired. Besides Dryander, there were two others who acted as sub-librarians ; and Dr. Brown, the botanist, who had the charge of the botanical collection. . . . At the time of which I am speaking, he might be seen daily in Sir Joseph's library, dissecting plants, and accumulating those stores of knowledge which have since gained for him the reputation of being the first botanist and botanical physiologist in the world, and the honour of being one of the very limited number of foreign Associates of the Academy of Sciences of Paris. . . .

". . . The attention which Sir Joseph Banks paid to the affairs of the Royal Society was unremitting. He was very much of an autocrat, but, like other successful autocrats, he maintained his authority by consulting the feelings and opinions of others, and no one complained of it. There is no doubt that his ample fortune, and his devotion of it to purposes of natural science, made his task more easy than it would have been otherwise.

Still, he could not have accomplished what he did if he had not possessed a great knowledge of human nature. . . . On the whole, it is difficult to conceive that any one could perform his duties of the Royal Society in a manner more honourable to himself, or more beneficial to the community, than that in which they were performed by Sir Joseph Banks." [1]

[1] Sir Benjamin C. Brodie : *Autobiography*, pp. 67–74.

CHAPTER XIX

SOME FRIENDS OF LATER YEARS

ONE of the closest friends of Banks was William Marsden, the orientalist, some ten years his junior, and F.R.S. in 1783. Marsden was in the service of the East India Company, in the island of Sumatra, from 1771–9. He was a man who could never be idle. He studied everything that came before his notice, including the difficult Malay tongue, and the people and resources of Sumatra. He was really happy in Sumatra ; but news came to him of the attention paid to travellers, and the spirit of curiosity excited, among the learned classes at home. The publication of Cook's voyages quite aroused him. In his own words, " the contemplation of these circumstances raised in my breast a longing desire to be allowed the opportunity of associating with such men, and to become a participator in their liberal pursuits." At first his plan was to devote himself to the literary life, satisfied with his small income. But he found it would suit him better to have a really active life, if he were to take the place in the world that he coveted, and became an East India Agent, with head-quarters in Gower Street. In 1781 he was offered an introduction to Sir Joseph Banks (by Captain Thomas Forrest, noted for his voyage to New Guinea).

" On March 1 I made my appearance at his breakfast table, where I found an assemblage of about a dozen persons eminent in Science and different branches of knowledge. Among them were Dr. Solander, Dr. Maskelyne, Mr. Dalrymple, Major Rennell, Dr. Blagden, Dr. Herschell, Mr. Planta, and others. My reception was so kind and encouraging as to remove at once all feeling of constraint or sense of my own inferiority; and I was led to take an active part in a conversation, to me of an interesting nature. . . . My desire of furnishing some account of the island in which I had resided was strongly encouraged—Sumatra being, as they observed, of all accessible places in the world, that which was least known. After taking leave, I received from the President a cordial invitation to meet him and his friends whenever it might suit my convenience, and to make free use of his books either in the library or at home. Until 1795, when I ceased to have the command of my own time, the rooms in Soho Square were my habitual place of resort, where I met a variety of persons, and acquired information of what was going forward in the world of Literature and Science."

Marsden's *History of Sumatra* was published in 1782. This book gave him a reputation which adheres to his name to this very day. It established his position in the society he had chosen, and gave him credit for a general ability which ultimately found resource in official life. In 1795 he entered the Board of Admiralty and later became Principal Secretary, until ill-health obliged him to resign his post in 1807. For another quarter of a century Mr. Marsden led a life of quiet happiness, one of the leaders of cultured society. His association with Sir Joseph Banks was of the closest and most friendly character, as is evidenced by the flying short epistles to one another on all conceivable topics. Often was Marsden requested to take Banks's vacant place at the Royal Society.

The successor of Marsden as Secretary to the Board of Admiralty was John Barrow, another product of that stirring Age; when rewards of the highest distinction came to men who knew how to seize the peculiar advantages which it offered to serious intelligence and industry. Barrow began life in a modest way as a clerk ; went to sea in a Greenland whaler ; taught mathematics at Woolwich ; was " discovered " by Sir George L. Staunton, who took him into China in the suite of Lord Macartney ; and had two years as his Lordship's private secretary at the Cape of Good Hope. All Barrow's leisured time, wherever he happened to be, was devoted to scientific and literary inquiry. He was an excellent botanist and geographer. When he was appointed to the second secretaryship of the Admiralty, in 1804, he had drawn the public notice by a published account of his travels in South Africa, and by his appearance in Society as a young man of varied attainments and general promise. He became a F.R.S. in 1806.

When Barrow became acquainted with Sir Joseph is not clear. Probably soon after his coming to London, through Sir George L. Staunton. It was not long before they were the best of friends. They had most ideas in common ; and Barrow was exactly the style of young man likely to commend himself to Banks. His official post brought them necessarily into communication ; and, as he held that post for about forty years, while he continued to hold a prominent place in the learned and scientific world, Barrow's recollections of his old friend may be read with great point and authority :[1]

" A similarity of tastes, my having been also a traveller, and my unfeigned respect for his character, soon established for me an intimacy with Sir Joseph Banks ; who invited me, with more than common cordiality, to join his Sunday evening conversations at his house in Soho Square. This intimacy continued without interruption to the last days of his life. . . ."

[1] v. *Sketches of the Royal Society and the Royal Society Club* (London, 1849).

[About 1815 he proposed to Barrow to be on the Council of the Royal Society. He was forthwith elected at the session of that year.]

" . . . When he became President, the general reputation in which Mr. Banks was held: the fame of his voyages, his wealth, and his liberal and courteous reception of all, and more especially of foreign visitors, gave him an extensive social influence, and made his house an agreeable centre of the literary world. In truth, the Chair of the Royal Society does not require to be perpetually and exclusively filled by men of science, or by persons elevated in any one particular department of Science. The President should be conversant in general knowledge, especially in knowledge of the world, courteous and agreeable in his manners and conversation, ready to oblige and to forward to the best of his power the objects brought to the consideration of the Society; in short, to follow the example of Sir Joseph Banks, in promoting intercourse among the members at certain fixed times set apart for that purpose. . . ."

" I think Sir Humphry Davy has not done justice to Banks in the character which he has drawn : ' a tolerable botanist,' ' a lover of gross flattery,' ' a house like a Court,' are expressions in my opinion unfounded and unjust."

" In spite of his physical infirmities, Banks was rarely compelled to forego appearing at his Sunday evening assemblies. Every new discovery in the range of Natural History, every new and ingenious specimen of art, every curious and useful invention was sure at some one or other of these meetings, to be exhibited. . . . Sir Joseph's house was itself a repository of arts, science, and literature. I had a general invitation to Spring Grove, consisting of woods and a small garden laid out with ornamental shrubs and flower-beds, and neatly kept under the inspection of Lady and Miss Banks. The ladies were most agreeable companions, and Sir Joseph was cheerful and much at his ease, which was not always the case when in town."

Barrow's writings on the great voyagers were of high interest to Banks. Sir Joseph writes : " You must come and dine with us some day to meet old Scoresby, the celebrated whale-fisher, who has given me more information about the ice in those regions than any other I have conversed with." Barrow had the satisfaction of filling this engagement to meet Captain Scoresby.[1] He was ardent to co-operate with Banks in promoting Captain Parry's voyage to the Arctic Seas. As to Geographical research generally, the mantle of Banks was now being shifted to the shoulders of Barrow, who, some few years later, took a leading part in the founding of the Royal Geographical Society.

The faithful Barrow's personal attention to Banks in these latest years was almost filial. He was Banks's right-hand man in the various scientific duties which continued in operation until the very last. The affectionate references to him which have just been quoted are the very best testimony to Banks's power to attract such men.

William Jackson Hooker was a rather notable acquisition to Banks's coterie during his later years. Their friendship appears to have begun about the year 1806. But as Hooker lived at Halesworth, partner in a brewery concern, he could not be part of the London circle of naturalists. The excursion to Iceland brought him very closely in association with Banks. After this, they had much in common as long as Banks lived.

When the *Tour in Iceland* was being prepared for the printer, Banks lent him some memoranda (probably the

---

[1] Scoresby speaks warmly of Banks's kindness and assistance ; including the loan of valuable instruments for submarine work (*Arctic Regions*, vol. I).

entire manuscript of his own Journal). The following note is curious as indicating his reserve as to publicity in some matters, especially arising from a consciousness of his want of literary power. His perennial generosity and readiness to help others is again very obvious.

" Soho Square, *June* 16, 1810.

" . . . The papers containing memoranda of some matters I saw in Iceland you are welcome to use as remembrances. But as I declined to publish them myself, you will easily understand that I had no idea when I put them into your hands of their being printed in my name, or with any kind of reference to me. As far as they may be useful to your own work they are entirely at your service."

Hooker protested that he could not see how to use these notes without saying to whom he was indebted for them. Whereupon Sir Joseph: " I see no objection to your using any part of the information as coming from me, when it is of a nature that renders the disclosure of its source necessary. It will be the more agreeable to me the more it is done in notes, where my name must come forward."

Botany and Brewing turned out to be occupations incompatible with one another, if great zeal was to be devoted to either. Hooker was not a good man of business; his love of botany kept him in thrall. He soon came to have a longing for travel in the tropics. The fascinations of plant-collecting seemed higher than those of money-making. As soon as Sir Joseph Banks learnt this inclination, he endeavoured to find for him an opportunity for its indulgence. His other friends, generally, wished him to stay at home, where there was plenty of work to be done in botany without his health being endangered in malarious deserts.

Dr. Horsfield was now home from Java and Sir Stamford Raffles, keeper of the India House Museum, and in touch with Sir Joseph. Hooker very soon had the opportunity he wanted, with all the advantages of the patronage of the Board of Trade and Sir Joseph Banks. Arrangements had not proceeded far, however, when fresh accounts of the horrible climate of some parts of Java, and a reminder of the sad accounts of Cook and of Banks himself, together with the renewed entreaties of his parents, prevailed upon him to renounce his plans.

Sir Joseph was vexed at all this. He had been at some trouble to arouse the authorities, through Lord Bathurst, on the importance of the project, and to arrange that Mr. Hooker should have good remuneration, and every advantage in his favour. The tradition is that he used very ungracious language to Hooker. There is one memorial of the affair[1] which seems to represent Banks's last endeavour to have his own way ; for that is what it comes to. He was now never thwarted by anybody, and was getting more irascible every year at the mere breath of such a thing. The letter now presented is too sarcastic for the occasion, but Banks's remonstrance shows plainly that it was a subdued exhibition of anger.

Sir Joseph Banks to W. J. Hooker.

"SPRING GROVE, *June* 19, 1813.

"MY DEAR SIR,—Though I really cannot think it possible that your relatives and friends in Norfolk can consider an island half as large as England to be of a deadly and unwholesome nature because one town upon it is notoriously so, I see their objections are urged with so much determination and eagerness that I am far indeed from advising you to despise them. I have, however, no

[1] Preserved among the Hooker correspondence at Kew.

doubt that arguments or injunctions equally strong will be urged by them if you attempt to extend your views further than the exhausted Azores, originally scarce worth the notice of a Botanist and now almost entirely transferred to Kew Gardens by the indefatigable Masson.

"From the complexion of all that has passed of late in the conduct of your friends, I have no doubt that they wish to force you to adopt the advice of Sardanapalus : to eat, drink, and propagate. How you will like to be married and settled in the country, as Joe Miller wish'd the dog had been who flew at him and bit him, I know not ; but that this fate is prepared for you somewhat earlier than the natural period of renouncing an active life is a matter of which I have no doubt. But, pressed as you are, I advise you to submit and sacrifice if you can your wish for travelling to the importunities of those who think they can guide you, to a more serene, quiet, calm, sober mode of slumbering away life, than that you proposed for yourself.

"Let me hear from you how you feel inclined to prefer ease and indulgence to hardship and activity. I was about twenty-three when I began my peregrinations. You are somewhat older ; but you may be assured that if I had listened to a multitude of voices that were raised to dissuade me from my enterprise, I should have been now a quiet country gentleman ignorant of a number of matters I am now acquainted with, and probably have attained to no higher rank in life than that of country Justice of the Peace."

But there was no permanent ill-humour between these two worthy men. Hooker was but twenty-eight years of age, and the most promising of the younger botanists. He was a regular visitor at Spring Grove, when he was in town ; just the man to attract Banks. Amiable and intelligent, and an athlete over six feet in stature, his personal qualities were significant of one who would make his mark in the world.

One day, at Banks's house, he met Dr. Abel, together with Thomas Manning, safely returned from his adventure in China and Tibet. The delights of this *rencontre* may be imagined. But Hooker had given up longing for such examples to copy. He had found abundance of material for study at home, and his first grand monograph, *The British Jungermanniæ*, had given him a reputation. About this time the brewery business was given up; and he asked Sir Joseph Banks if there was likely to be any botanical appointment within reach, by which he could improve his income. Banks promptly told him that the Professorship of Botany at Glasgow University was vacant, and he was prepared to use what influence he possessed to obtain it for him.[1] He told him there was a noble Botanical Garden at Glasgow, well endowed both by the University and the City, " toward the development of which Kew would place all its resources." Hooker accordingly went to Glasgow, and remained there with well-earned distinction for twenty years. In 1840 he accepted the Directorship of the Botanic Gardens at Kew, and held it till his death in 1865. He was knighted in 1836.

The ideal Director for Kew Gardens was found in W. J. Hooker. He entered upon his duties to find " an ill-kept garden of some eleven acres, with a few tumble-down houses." It was being neglected. There was even talk of abandoning the Institution altogether. The energy and the taste of the new Director speedily transformed its appearance and raised its value. At the time of his death the premises occupied two hundred and fifty acres. The Palm-House and the Temperate House were erected, together with three Museums. Beside all this, a separate building was provided for the Herbarium,

[1] *v.* notice of Sir W. J. Hooker, in *Annals of Botany*, vol. XVI.

and a botanical library in connection with it. This was
all a fitting result to the succession of Banks's latest disciple. His powers of organization, his entire mastery of
botanical science, and his ability to deal with mankind,
were universally acknowledged. More than this, his personality was charming. The eulogium of Asa Gray is
everything that one might wish for from his fellow-men :
" Of pleasing address, frank, cordial, and of a very genial
disposition. . . . None knew him but to love him,
none named him but to praise. . . . A model Christian
gentleman." [1]

Hooker's house at Halesworth attracted a number of
cultured people, especially those with a turn for botany.
It was in close touch with James Edward Smith and the
Norwich people.

One day, in 1816, Monsieur De Candolle appeared upon
the scene, and spent several days at Halesworth, which
were, he says, exceedingly agreeable. Young John Lindley
was there at the time as a pupil of Hooker. De Candolle
took home with him the best recollection of England and
her scientific men. His intercourse with Banks is best
described in his own words :

" Dr. Marcet me conduisit le matin chez Sir Joseph
Banks. C'était le but de mon voyage, et sa maison a été
celle où j'ai passé presque tout mon temps. Ce respectable

----

[1] If Sir William Hooker rescued Kew Gardens from neglect and
oblivion, it remained with his distinguished son and successor to bring
the institution to its present position of utility and grandeur. Sir
Joseph Hooker was born at Halesworth, in June, 1817. After several
years of travel, in company with the expedition of Captain J. C. Ross
to Australasia and the Antarctic Islands, he spent several years in
India, enriching botanical science with profuse details of the Himalayan
and Indian floras. He joined his father at Kew, became Assistant-
Director of the Royal Botanic Gardens, and succeeded to the Directorship in 1855. Since his retirement from the post in 1885 he has
always been in touch with Kew. As late as the summer of 1910, this
Grand Old Man of science was still to be seen at the Herbarium on
botanical research. This in his ninety-fourth year ! Sir Joseph was
President of the Royal Society 1873-8.

vieillard, alors perclus par les suite de la goutte, se faisait rouler dans une fauteuil jusqu'à l'une des salles de son Herbier, où les habitués venaient causer avec lui. J'étais malheureusement mal placé sur ce rapport, par mon ignorance de l'Anglais. M. Banks ne parlait pas français, de sorte que nous étions réduits à nous dire quelques mots isolés, ou à nous servir d'interprétes. Malgré cette contrariété, je puis dire que j'étais assez bien avec lui. Il m'invita à sa jolie campagne de Spring Grove, à ses soirées du Dimanche. . . . Il mit son Herbier et sa bibliothèque à ma disposition. . . . Je venais chaque matin travailler de dix à quatre heures, et outre les ressources immenses que me présentait ce musée botanique, j'en profitai beaucoup pour connaître les botanistes anglais qui s'y rencontraient fréquemment."

To complete this picture, it should be mentioned that Robert Brown was the rival of De Candolle as " the first botanist of Europe." They were very happy together in Banks's house.

Humphry Davy was one of the new generation which clustered round Banks in his later years. He was often to be seen in association with the President, who had been one of the first to detect Davy's signal merit. He made his appearance as a lecturer at the Royal Institution at a very early date, and doubtless contributed to the success of that body.

Davy left behind him a notice of Banks, very freely drawn : not ill-naturedly, but alluding to certain defects which others had noticed with some disposition to caricature. " He was a good-humoured and liberal man, free and various in conversational power, a tolerable botanist, and generally acquainted with Natural History. He had not much reading, and no profound information. He was always ready to promote the objects of men of science, but he required to be regarded as a patron, and readily swallowed gross flattery. When he gave anecdotes of his voyages he was very entertaining and unaffected. A courtier in character, he was a warm friend to a good King. In his relations to the Royal Society he was too personal, and made his house a circle too like a Court."

Some of these statements might with justice be disputed ; but controversy is needless upon the subject.

After a temporary chairmanship of the Royal Society by Dr. Wollaston, Davy succeeded to the Presidential seat, and held it until 1827. He found it incumbent upon him, like his predecessor, to keep up some appearance of State, and occupied the chair in full Court-dress. He continued Banks's practice in weekly receptions, changing the evening from Sunday to Saturday. " He had to put up with the annoyances caused by hurting the feelings of rejected Fellows."[1]

According to Dr. Paris, Sir Humphry Davy " sought for the homage due to patrician distinction," and thereby incurred some unpopularity.[2]

Of other rising young men who enjoyed Banks's intimacy near the end of his life, several should be mentioned who afterward rose to great distinction.

William Edward Parry was introduced to Banks in the year 1818, by his friend Barrow. The young Lieutenant availed himself gratefully of the opportunities thus offered, by making free use of Banks's library, and attending his social functions. These, he remarks, were " not like those of fashionable life, but given from a real desire to do everything which could, in the smallest degree, tend to the advancement of every branch of science."

Charles Waterton speaks very highly of Banks's attentions to him, and his appreciation of Waterton's services to Natural History. Some portions of his manu-

[1] *v. Memoir* by his brother Dr. John Davy.
[2] *Life*, by Dr. J. A. Paris.

script journal were made of use to the Tuckey Congo Expedition.

Waterton's final allusion to Banks is an echo of the popular opinion of his time. " Death robbed England of one of her most valuable subjects, and deprived the Royal Society of its brightest ornament."[1]

William Buckland (afterward Dean of Westminster), with his friend William Conybeare (fellow-geologist and afterward Dean of Llandaff), became much attached to Banks in his old age. Conybeare was one of the earlier members of the Geological Society and F.R.S. in 1819. Buckland had joined the Royal Society the previous year. He had already shown great promise of his future distinction in science, and was excellent company for Banks. After the latter's death he planted at Christ-church and at Islip a quantity of *Rosa Banksiæ*,[2] as a token of regard for his memory.[3]

Henry Brougham had been a friend of Banks from his youth. He joined the Royal Society in 1803. He always had a smattering of science, but he was not a devotee of anything but his own self-advancement. As early as December, 1800, he was " disgusted with " the legal profession, and was " resolved to attempt an opening in the political line " ; and begs that Banks will remember him if a chance opens for him. When the *Edinburgh Review* began to startle the world, Banks appears to have taken exception to its tone ; and, besides, pointed out some inaccuracies in the critical references to the African Association and other matters in which he was interested. Brougham replied to this in apologetic vein ;

[1] *v.* Pref. to *Wanderings*.
[2] Or Lady Banks Rose, so named by Robert Brown ; originally sent from Canton in 1807 by William Kerr.
[3] In Buckland's *Life* there is inserted a sketch of his own, of Banks's library : with the helpless old gentleman propped up in a deep chair ; Brown is at the shelves, Buckland and Conybeare are bidding adieu. With a little idealization, it would make a capital subject for a portrait picture.

but the *Edinburgh* was selling splendidly ("the first edition in 5 days, the second going off rapidly "), and Brougham was already in full sail upon what turned out to be a really memorable career.   Henceforth, there was a distinct gulf between them on public matters ; but not wide enough to break their friendship.

Brougham's memorial of his old friend, in the form of an essay upon his life and doings and his services to the country, is proof of the very high regard and affection he had for Banks.   Apart from the necessary rhetoric, its pages breathe the truest personal sentiment toward him, while they record the achievements of " a life devoted to the love of wisdom."

CHAPTER XX

"A FINE OLD ENGLISH GENTLEMAN"

Sir Joseph Banks to Sir Everard Home.

"SPRING GROVE, *September* 24, 1817.

YOUR permission to visit Lincolnshire this autumn gives me spirits to undertake the journey. I felt something so near the necessity of going, that it would have been a want of courage to decline it,—which my countrymen would have censured, had I passed the autumn without a return of my last year's complaint. I have arranged my journey with very short stages; none so much as forty miles a day, and on three of the four days I must spend on the road I shall be received at the houses of my friends. . . . On Sunday I start, and stop the night at Lord Salisbury's at Hatfield. The first week in November brings me back, so that in truth I shall have little time to grow worse. I expect every minute the arrival of the Queen and Princesses from Windsor, who have promised me the honour of a visit this day."

When Banks wrote this letter, he was recovering from an attack of gout. Sir Everard was himself an invalid, with threats of the same tiresome complaint as his friend. Banks evidently meant to go to Revesby once more. It turned out to be the very last time he was able to do so. He writes to Home on October 5, having just arrived:

" Our journey was very slow, not more than thirty-five

REVESBY ABBEY

miles a day, and we lodged three nights out of four at gentlemen's houses. I bore it without inconvenience, and am here as well at least as I was at Spring Grove. My ladies, who are both rather crazy, approve much of this leisurely mode of travelling. We shall therefore return in the same manner as we came ; and, I trust, dine in Soho Square on the Wednesday before the Society meets.

" Give my love to my god-son, and thank him for his kindness in writing to me so frequently when I was anxious on your account. In his last letter he apologizes for its not being longer, although it fills a page and a half. Pray assure him that the shorter a letter is which communicates what is to be transmitted, the better both for the writer and the reader. Women who have little to do scold each other for shortness in letters. Men who have anything to do know the value of brevity."

It would be unsatisfactory to close this history without at least a peep behind the scenes at Revesby Abbey. The tale would lack an important feature if Banks's doings as a country squire were entirely obscure. He was a thorough County man, experiencing to the full all the delights which belong to rural business and rural pleasures. His home life at Revesby exemplified the character in the famous old song :

> " Of a fine old English gentleman
> Who had an old estate,
> And who kept up his old mansion
> At a bountiful old rate. . . ."

The house at Revesby was a large, square, uniform building, dating from the latter part of the seventeenth century ; not altogether attractive in its somewhat cumbrous shape, but capacious and comfortable. The park, given over to sheep, cattle, and deer, situated on the last slope

x

at the southern extremity of the Wolds, gave to the house
a grand view of wood and the illimitable pastures of the
fen-land. The pretty village, ranged on two sides of an
unusually spacious green, was famous for its cricket-
matches; and for a yearly Fair which drew the country-
folk in great numbers from villages near and distant.
The parish church was in the classic or Italian style,
dating from 1735.[1]

Sir Joseph Banks really loved his country home. From
August to the end of October in every year he revelled
in farming and gardening, and in country pursuits and
pleasures. He always attended the village Fair, and
with some state ; bringing with him a lively house-party.
The Squire would drive round and round again, accompanied by his Lady, buying generously at the booths,
and distributing his purchases among the visitors who
flocked thither in their hundreds. This was one of the
most popular Fairs in the county. At Revesby it was
a day of real old-fashioned festivity, when Mummers and
Morris dancers vied with up-to-date entertainers in
making the most of a joyous day.

Sir Joseph Banks to Mr. Tyssen.

"LINCOLN, *September* 11, 1805.

" My DEAR SIR,—The ladies beg me to assure you and
your sister that we shall be happy to see you both at
Revesby at the time you have appointed ; and we all
hope that you will manage your affairs so as to let us
have as much of your company as you can spare. . . .
The town is full of gaieties here. We have two Races and
an Assembly for every day, and gentlemen who ride their

---

[1] A handsome new Decorated church was built by the Rt. Hon.
Edward Stanhope, in 1891. The old mansion was destroyed, and the
present fine Elizabethan building raised in its place by Mr. J. Banks
Stanhope, in 1849.

own horses ; the whole to end in a grand breakfast and ball on Saturday at Lord Monson's, four miles off. We should have been glad to have the pleasure of your sister's company to partake of these festivities, which suit the ladies so much better than they do either you or me. Give me a line to tell me whether I can be of any use to you by ordering horses at Boston, or sending my chaise. . . ."

Among the house amusements at Revesby few were in such favour as Archery. While Banks was vigorous enough to take part in it, he maintained a very high level in the scoring. His sister was not far in arrear as an accomplished toxophilite. Mistress Banks kept the score for a long series of years. The cards, in her beautifully clear handwriting, are now preserved in the British Museum.[1]

Angling was another enjoyment of Banks's leisure, which he had practised from his earliest years. The time came when that leisure was encroached upon ; but he could sometimes snatch a few days for sport. He tells Sir William Hamilton of his doings at home. On one occasion he writes : " I am here buried among sheep, wool, etc. ; so much that my favourite amusements of fishing and shooting surround me and I scarce ever taste them. . . ." Again he says : " We drew ten miles of fresh water, and in four days caught seventeen hundredweight of fish ; dining always from twenty to thirty masters and mistresses, with servants and attendants, on the fish we had caught, dressed at fires made on the bank ; and when we had done we had not ten pound of fish left." [2]

---

[1] Addl. MSS., 34721.
[2] Arthur Young has a note of Banks's piscatorial feats. ". . . I found these fishing-parties, which lasted over four days, spoken of by many persons with great pleasure. Miss Banks has kept a particular journal of these piscatory excursions, which is decorated with many

A great feature of Revesby life was the garden. Here were undertaken many of Sir Joseph's planting experiments ; and here were deposited a vast number of the curious and elegant exotics which constantly found their way into his hands. To this day the garden of Revesby Abbey is a paradise. Beautiful trees from every country on the globe, flowers of infinite variety, meet the eye at every turn. The scene is one that could only have been devised and perfected by consummate taste, allied to a thorough knowledge of the resources of floral grandeur.

Several hundred deer were kept in the park. Banks was once induced to compute the profits accruing from deer-keeping, and found himself usually a gainer. The large area of 340 acres supported numbers of sheep and cattle besides. The spacious adjacent fen-lands were undergoing, in Banks's time, a continuous process of enclosure ; and these became famous grazing lands. During the active part of his lifetime the entire county was fired by his example, both as landlord and agriculturist. His advice and his opinions, on crops, and stock, and farming generally, were eagerly sought.[1]

Banks had been long associated with Sir John Sinclair and Mr. Arthur Young in the improvement of agricultural methods. Young was wonderfully helped by really competent persons. The hospitality of the best houses all over the country was opened to him.

Sir Joseph Banks was soon hand-in-hand with Arthur

drawings. She had the goodness to favour me with the following totals :—

In 1788, 1764 lbs.	In 1791, 842 lbs.	In 1794, 1366 lbs.
,, 1789, 693 ,,	,, 1792, 1410 ,,	,, 1795, 2567 ,,
,, 1790, 1711 ,,	,, 1793, 2644 ,,	,, 1796, 1562 ,,

In Sir Joseph Banks's kitchen is the picture of a pike that weighed thirty-one pounds, which was thirteen years old." (*Agricultural Survey of Lincolnshire*, p. 395.)

[1] ". . . We are grown proud of heart. . . . I never saw my neighbours in so high spirits. The Fens are drained and peopled, and agricultural industry is active among us " (Banks to Marsden, 4 Oct., 1810).

Young. A bulky collection of letters [1] testifies to the intimacy of their intercourse, and to the zeal with which Banks could escape from the multiplicity of other interests, to thoughts about Wool and Sheep. At this time (1786–7) legislation on the exportation of wool was impending. But for several years past there had been much experiment and inquiry on the improvement of breeds of sheep, in which Banks had taken some part, and he had become a capable authority. Hence, the President of the Royal Society had one more responsibility thrown upon his shoulders ; and it appears, from every quarter, that he came to have a very well-deserved reputation for a practical judgment on rural affairs.

In the course of his tours in England and Ireland, Arthur Young visited all the best estates ; and, in process of time, gave to the public an ample record of his experiences. Of the County Surveys, which were prepared under his superintendence, that for Lincolnshire is fortunately by his own hand,[2] since we are enabled to get into actual touch with Sir Joseph's own estate.

Those were, indeed, palmy days for Agriculture. It was estimated that about half of Banks's property in Lincolnshire brought in rentals amounting to £5721, from 268 tenants. In the manor of Revesby there were 62 farms for the rental of £1397, upon 3401 acres. "This vast division of farms arises from a determination in Sir Joseph Banks not to distress the people by throwing them together, by which he loses much in rental, and sees a property ill-cultivated ; and which must be the case till by deaths he can gradually but very slowly improve it. . . . Sir Joseph has no objection to granting leases, but he is never asked for them. Seeing a tenant improving his

---

[1] Addl. MSS., 35126–31.
[2] The [British Museum copy (988, g. 9) was formerly in Banks's library ; presented by the author with his " respectful gratitude for so much valuable information relative to Lincolnshire as contributed very materially to the facts contained in this work."

land by hollow draining, he gave him a lease of twenty-one years, as a reward and an encouragement." [1]

During the year 1787–8, there was a lively intercourse between Banks and Young ; arising from proposed legislation on the subject of Wool. The manufacturers were making a vigorous attack on the growers of wool. A Bill was brought into Parliament, prohibiting its export and placing unwelcome restrictions on its sale with a view to hinder its being smuggled into France. The great proprietors were naturally up in arms at this appearance of an attack on the landed interest. A representative Committee was formed to oppose the Bill, by deputation from the counties. Lincoln sent up Sir Joseph Banks, and Suffolk Mr. Arthur Young. By a good deal of energy and plain-speaking this committee secured some modification of the features of the Bill. But it passed into law (May, 1788).

Sir Joseph Banks to Arthur Young.

" SOHO SQUARE, *May* 13, 1788.

". . . I give you joy sincerely at the glory of being burned in effigy. Nothing is so conclusive a proof of your possessing the best of the argument. No one was ever burned if he was in the wrong . . . when argument is precluded firebrands are ready substitutes."

At the beginning of the nineteenth century there were more sheep, and better sheep, in Lincolnshire than in any county in England. The breed was reckoned to produce the very best wool, a circumstance due to the introduction of merinos. In point of fact, the whole countryside was raised in value by the experiments and improvements of recent years, to which Sir Joseph Banks had given his

[1] Young : *Agriculture of Lincolnshire* (London, 1799).

untiring attention. The following strong sentence from a letter of Lord Liverpool speaks volumes for the reputation which Banks's zeal had shed abroad among people able to gauge its worth :

(October 25, 1804.) ". . . I rejoice to find that you are able to attend the business of your county in cases where your advice and influence are of so much importance. The great prices at which the lands brought into cultivation by measures which you recommended have sold, are a proof of the value of those lands, of the wealth of the county, and of the wisdom of the means employed for enclosing, draining, and otherwise improving them. . . ."[1]

It was about the year 1787 when George III took into consideration the improvement of the breed of sheep. He gave orders for the importation of a small flock of Spanish merinos, and became an attentive observer of breeding operations. With the help of Sir Joseph Banks, his farm near Windsor was managed on the improved principles then coming into vogue. After a dozen years or so the movement took very large proportions. By sale, and by personal gift, the new breed (of Spanish with the best British) was getting represented everywhere in Great Britain. After 1804, an annual sale of the King's flocks was held at Windsor, when prices ruled very high.

[1] An excellent account of the reclamation of the adjacent fen-land is given by Smiles in his *Life of John Rennie*. Of Sir Joseph and his energetic share in the project he writes very warmly. "Farther : he was a popular and well-known man, jolly and good-humoured, full of public spirit ; and, though a philosopher, not above taking part in the sports and festivities of the neighbourhood in which he resided. . . . From an early period Sir Joseph Banks entertained the design of carrying out the drainage of the extensive fen-lands lying spread out beneath his hall window ; and making them if possible a source of profit to the owners, as well as of greater comfort and better subsistence for the population. The reclamation of these unhealthy wastes became quite a hobby with him ; and when he could lay hold of any agricultural improver, he would not let him go until he had dragged him through the Fens, exhibited what they were, and demonstrated what fertile lands they might be made. . . . His county neighbours were very slow to act, but they gradually became infected by his example, and his irresistible energy carried them along with him."

The interest in this affair never flagged while George III was able to have a personal share in it.

When the King's last illness was confirmed, Sir Joseph began to relinquish his own occupation as a farmer. It is easy to understand that he was now become out of touch with it, now that the King was no longer in touch with him. After so long a period of collaboration in utilitarian matters of all sorts, it is not unlikely that life was losing one of its first incitements. Yet, Banks's wonderful natural energy prevailed. In some respects, the years of his life after the sixtieth were not less devotedly at his country's service, and hardly less efficient ; as we have seen. For one thing, the improvement of sheep appears not to have reached anything like finality. It must needs go on. He could not cease to be a farmer, after all. Some time in 1811 a Merino Society was started in London, with Sir Joseph as President. How long it lasted is not easy to say. But for two or three years Reports of its doings were issued to the public.

The number of local interests in which Banks had a share is astonishing. He devoted himself to Lincolnshire while he was at home. The Fens were drained, a process extending over many years. The Horncastle Canal was constructed ; the Witham navigation was improved ; Boston harbour was deepened and renovated, and brought into line with the needs of a busy port. At Louth a woollen manufactory was established, with the view of utilizing Lincolnshire wool only. In fact, the whole county felt the influence of Banks's energy and his liberal views concerning the public weal. As all this went on for a period of nearly sixty years, the people of Lincolnshire had an unusual regard for their leading Squire, and held his name in veneration long after his death. Happily, his successors at Revesby have kept up the tradition of his good work.

The people of Horncastle keep his memory green.

Banks's estates included this parish; and, along with Revesby and the surrounding villages, rejoiced in their connection with a model landlord. The Public Dispensary in the town was established in 1789 with Banks's liberal aid. He was made President of the Institution, and the value of his support was more than once acknowledged in its minutes. In point of fact, Banks was beloved at Horncastle and everywhere around. One of the anniversary meetings of this Dispensary (1813) was actually postponed because his health at the moment prevented his personal attendance.

Banks had also friends and associations at Spalding. Even such a matter as the remuneration of the postmistress there is referred to him. He was a member of the Spalding Gentlemen's Society as far back as 1768, and several times gave books to the library. At Boston, various public improvements were carried through under his patronage. There is a fine portrait of Banks (by T. Phillips) in the Guildhall, which was presented by him when he became Recorder of the borough in 1804. The people of Boston made more than one endeavour to induce Sir Joseph to stand as their representative in Parliament, but without avail. It was useless to tempt him in that direction. Looking backward upon his record, one cannot but agree in the wisdom of this decision.

Sir Joseph's attachment to his own county is further signalized by the assistance or the patronage of his neighbours, when he could serve them. Captain Flinders, who came from Dorrington, owed some of the notice he got from the Admiralty to the fact that he was a " countryman " of Banks. Young Franklin, born at Spilsby, and a relative of Flinders, doubtless occupied a niche in the mind of Banks due to the fact of his coming from Lincolnshire. Among several instances of this spirit which could be given, there is a pleasant one related by William Fowler, an architect and builder of Winterton : " After a very agreeable and familiar conversation he says to me, '*My countryman*, I will undertake your business for you, and make the best enquiry I can. . . .'" [1]

Nor was Sir Joseph less zealous in archæological studies of his county. As far back as the year 1786 he was collecting materials for a history of Lincolnshire. Dr. Hunter, of York, appears to have been helping in the matter. Richard Gough also was associated in the project. But county histories are more easily talked of than accomplished ; and all that remains of this effort has got into the custody of the British Museum. If, however, he found this undertaking impossible with so many calls upon his time and energies, he was always on the alert if any relic of old times was unexpectedly revealed, or some effort was to be made to preserve the precious heritages of the past. A difference of opinion with the Dean and Chapter of Lincoln supplies an amusing instance of the heat which Banks could put into a question when he felt strongly on the point at issue.

The twin spires which formerly surmounted the western steeples of the Cathedral were demolished, in the year 1807, on the ground that the appearance of the building would be thereby improved. There was a good deal of local indignation over the matter. Naturally, one side supported the Cathedral authorities ; the other side deemed it an act of sacrilege. Sir Joseph proposed an Association for the purpose of " compelling those persons " to repair and restore the spires, if a sufficient fund could be raised. He caused a small copper plate to be engraved, of the Cathedral standing with two shadowy ghosts of the departed spires. He further committed himself to

---

[1] *Notes on William Fowler* (Hull, 1869). This Mr. Fowler was worth notice. He made some fine drawings of mosaics and stained-glass windows, which were submitted to the Society of Antiquaries. Banks remarked of them, " Others have shewn us what they thought these remains ought to be, but Fowler has shewn us what they are : and this is what we want."

THE ANTIQUARIAN SOCIETY.

SIR J. BANKS   DR. DIBDIN   SIR H. ENGLEFIELD   LORD ABERDEEN   MR. LYSONS   LORD MULGRAVE (PHIPPS)

MR. FLAXMAN   DUKE OF NORFOLK

THE ANTIQUARIAN SOCIETY
From a caricature by J. Gillray

some doggerel verses, and had them printed for circula-
tion—

> "Adieu, ye twin sisters, fair spires
> By learn'd architects anciently rais'd," etc. etc.

He went also to the expense of legal opinion on the
matter, from Doctors' Commons. The result was dead
against the right of any inhabitant or landholder in the
diocese to question the discretion of the Dean and Chapter, or to compel them to account for their proceedings.
So this agitation was speedily at an end.

Banks's lifelong connection with the Society of Antiquaries has been mentioned. Several communications
were made by him to the Society concerning local excavations, and some instances of objects being unearthed by
accident.[1] He was immediately interested by any discovery of the kind, and made it his business to have it
properly recorded. One (unpublished) memorandum of
his own relates to the three barrows near Revesby village,
which had been noticed by Stukeley in *Itineraria
Curiosa*, but not closely examined. In October, 1780,
Banks employed two men to open one of these barrows.
The result showed that the heaps could not be very
ancient. At the depth of about eleven feet, were pieces
of a small twig, or switch, not yet decayed. He deposited
a stone before covering in again, bearing date 1780, so
as to certify to later explorers that it had been opened.

Two days later, some examination was made in the
neighbourhood of the Abbey ruins. As far as they went,
however, there was little to reward the labours of the explorers.

The remains, such as they are, of the ancient monastic
ruins, are just south of Revesby village, near the edge of
the great Fen. In 1811 there were existing " foundations
of walls, and a fragment two feet high of brick and stone,

---

[1] *v. Index to the first fifty volumes of Archæologia.*

with loose stones here and there. . . . On the south, within the area, are mounds of earth, like barrows."[1] Of course, the materials of the Abbey buildings had gone in the usual manner : the stones were utilized by the second Joseph Banks, when rebuilding the parish church.

Banks's methodical habits were in active operation at Revesby. Domestic matters appear to have been ruled in the same spirit as his official concerns ; accounts were kept, and records made, in a way creditable to the best man-of-business. There are some scanty evidences, in his papers, that he could always give a remote date, or a price, or the gist of a conversation, with perfect accuracy. Both at Soho Square, and at his Lincolnshire home, books, papers, knick-knacks, curiosities, botanical preparations, were kept in the most absolute order.[2]

The house of Spring Grove, in Heston, was taken by Banks in the year 1779. There was a good old garden then in existence ; with a strawberry-bed of seventy-five feet in length, and corresponding amplitude for vegetables and fruits sufficient for a well-kept house. The new occupant soon developed it, upon the lines appropriate to an importer of exotic plants ; and the garden became cele-

---

[1] *Gentleman's Magazine*, 81, i. 19. In the year 1869 excavations were made on this site by the Rev. Thomas Barker, incumbent of the parish. *v. Associated Archæological Societies' Journal*, X, 22–25.

[2] Thanks to Arthur Young (*General View of the Agriculture of Lincoln*), we get one faint, but intelligible, glimpse inside Banks's house at Revesby : " His office of two rooms is contained in the space of thirty feet by sixteen. There is a bricked partition between, with an iron-plated door, so that the room in which a fire is always burning might be burned out without affecting the inner one. . . . Here were one hundred and fifty-six drawers, each with inside measurement of thirteen inches by ten, all of them numbered. . . . There is a catalogue of names and subjects, and a list of every paper in every drawer ; so that whether the enquiry concerned a man, or a drainage, or an enclosure, or a farm, or a wood, the request was scarcely named before a mass of information was before me. Fixed tables are before the windows (to the south) on which to spread maps, plans, etc., commodiously. . . . Such an apartment, and such apparatus, must be of incomparable use in the management of any great estate, or indeed of any considerable business."

brated for its high culture and for its curious and beautiful treasures from distant lands. The springtime brought numerous visitors to this house, being the period between Banks's residence in London and the ruralizing at Revesby. "Here he dined daily at four o'clock, in order that his frequent visitors from London might have ample time to return home in the evening. When the weather permitted, his guests adjourned to have tea and coffee under the cedars in the garden. In the intermediate time it was not unusual to visit his hot-houses and conservatories, under the auspices of his unmarried sister, Miss Banks ; or the dairy, under the especial care of Lady Banks, who was proud of displaying a magnificent collection of old china-ware which was there deposited." (B. C. Brodie, p. 74.)

When the Horticultural Society was set on foot, Sir Joseph was thus prepared to assist materially in carrying out its objects. Over twenty years of practical work and experiment had made him a high authority in fruit and flower culture ; and, as may be seen by the list of his published contributions to the Horticultural Society's *Transactions*, his knowledge and experience covered a wide field. By this time, Spring Grove was become famous for its horticultural wealth. Foreigners were made welcome, and Banks's intimate friends were often there by standing invitations.

Spring Grove is one of the pleasant places lost to the present generation, through suburban encroachment. Our picture of the house, as it appeared a century ago, suggests a real adjacent " grove," and possibilities of a large and roomy garden. The name is alleged by Banks to come from an actual spring in the woods, from which a water supply for the household was obtained ; and which also fed a convenient pond, thence escaping to Smallberry Green. The pond was the scene of various experiments. One of these was the raising of the American Cranberry, upon an artificial island ; a measure which met with great success, no less than one hundred and forty bottles of cranberries being the produce for 1813. Another was the growing of *Zizania aquatica*, a singular grass used for food by the Indians in Canada, from seeds imported in 1791. In a very few years' time this tender annual had benefited by its acclimatization, and was become a hardy and prolific plant, covering the pond quite thickly.[1]

It was in the improvement of apples, peaches, grapes, figs that Sir Joseph was most successful. Strawberries also were a great feature of Spring Grove : the original three hundred and seventy-five square feet of strawberry plants increased at last to five thousand six hundred and forty-five feet. It was Banks that popularized the system of mulching with straw, as above mentioned.

Many new importations of flowers are on record which were planted first in Spring Grove. The *Rosa Banksiæ* was sent by Wm. Kerr from China to Kew, and also to Banks's garden, where it became a great favourite, and much attention was paid to its cultivation. Isaac Oldaker (Lady Banks's gardener) submitted it to the Horticultural Society in 1820, remarking that " by care it had been transferred from an insignificant green-house plant into a hardy and splendid creeping shrub."

The Pæony was another triumph of these gardens. The first came from China in 1789, and was cultivated at Kew. In 1805 a splendid Double Scented Pæony (*P. albiflora fragrans*) was received by Banks, and added to the glories of Spring Grove.

The *Hydrangea hortensis* is understood to have been seen in England as early as 1740 (*Paxton*, p. 293). When

---

[1] " The ladies had a small pond stocked with gold-fish. One day Lady Banks told me that Sir Joseph had a visit from two young Americans. They were shewn round the grounds, and coming to the edge of the pond, were asked if they had any fish of the kind in their country. ' No, we have not,' said one. The other said, ' They appear to be a species of red-herring ; but I never heard of, and never saw till now, a red-herring alive.' "—Barrow : *Sketches*, etc.

introduced to Kew Gardens in 1789 (at first a plant with green petals which had flowered in the Custom House) it was a surprise and delight to every botanist. Banks made exhibition of it at his house in Soho Square, in the presence of a numerous company. The *Hydrangea* was carefully nurtured at Kew Gardens, established itself well, and became the parent of a numerous progeny.

Several of Banks's friends, inspired by his example, improved their gardens and built new hot-houses. One of these was Charles Greville, the philandering nephew of Sir William Hamilton, who had a fine garden at Paddington Green. He proved a good horticulturist in his later years, and helped in the importation of new exotics. Perhaps the closest rival to Kew, early in the nineteenth century, was the garden of Mrs. Joseph Marryat, at Wimbledon.

The traditions of Spring Grove were worthily maintained during the remainder of Lady Banks's lifetime.

Mr. Thomas Andrew Knight was associated with Banks in matters of horticulture. His brother Richard Payne Knight, equally distinguished in other walks, as virtuoso, collector, and arbiter in Ancient Art, was a particular crony of Banks. They were neighbours in Soho Square, and fellow-members of the Dilettanti Society. The Board of Agriculture desiring answers to a set of queries which were to be addressed to the best cultivators in the country, relied on Sir Joseph for his counsel and assistance. Upon Payne Knight being asked for the name of one in his neighbourhood, he mentioned his brother as more likely than any one else he knew capable of fulfilling the object in view. In point of fact, T. A. Knight was a pioneer. He anticipated Charles Darwin in assiduous experiments in hybridization and fertilization. But he lived a retired life, near Downton, in Herefordshire, far from the madding crowd. It might have happened that he would never have been heard of but for this timely association with Sir Joseph Banks. However, from this period (about 1791) their bond of friendship lasted as long as they lived ; to their mutual esteem, and to the steady progress of horticulture and fruit-farming. Knight became a copious correspondent. Each letter is a long and practical treatise ; and Banks must have sometimes felt a strain upon his powers as a pupil if he were to master the abounding details offered to him. But he was himself completely in earnest in recognizing the improvements necessary in fruit and flower culture; and it is evident from Knight's remarks that his correspondent was following his steps closely making the matter his own, and repeating his experiments.

These two men were the mainstay of the young Horticultural Society. Knight became the second President, and held the office for a long period of years. The early volumes of the *Transactions* bear witness to their industry and to the soundness of their views. Moreover, Knight's contributions to Vegetable Physiology keep their authority in modern text-books.[1]

---

[1] It has been shown that Sir Joseph abstained from authorship as far as the public were concerned. A few items only, on Agriculture and Horticulture, were printed under special circumstances.

The *Annals of Agriculture* published the following articles :—

A Report on Wool	Vol. IX, 288
Instructions given to the Council against the Wool Bill	,, - 480
Notes on Spinning	,, X, 217
A New Hay-barn, and New Rick-cloth . .	,, ,, 520
On The Hessian Fly	XI, 422
On the late Season, 1790	XV, 76
On the Hastings Turnip	,, 77
Account of Twelve Lincoln Sheep . . .	,, 357
On the *Musca pumilionis* (a fly attacking Rye) .	XVI, 176
Reply to Queries on Labour in Lincolnshire .	XIX, 187
Economy of a Park	XXXIX, 550

Papers printed in the Horticultural Society's *Transactions* :—

An attempt to ascertain the time when the Potatoe (*Solanum tuberosum*) was first introduced into the United Kingdom	Vol. I, p. 8
Some hints respecting the proper mode of inuring Tender Plants to our climate	,, 21

On the revival of an obsolete mode of managing
    Strawberries .   .   .   .   Vol. I., p. 54

An account of the method of cultivating the American
    Cranberry (*Vaccinium macrocarpum*) .   .   .   ,,   75

On the Horticultural Management of the Sweet or
    Spanish Chestnut-tree .   .   .   .   .   ,,   140

On the Forcing-houses of the Romans, with a list of
    Fruits cultivated by them now in our gardens .   ,,   147

A short account of a new Apple, called the Spring Grove
    Codling .   .   .   .   ,,   197

On ripening the second crop of Figs that grow on the
    new shoots .   .   .   .   .   .   ,,   252

Some Horticultural observations, selected from French
    authors .   .   .   .   Appendix

Y

CHAPTER XXI

THE END

I N his seventy-fourth year, Banks had one more carriage accident ; this time having a rather beneficial result than otherwise, for in the course of a restless night after, he voided a large stone and was consequently relieved from one source of pain.

Sir Joseph Banks to Sir E. Home.

" SPRING GROVE, *August* 27, 1818.

" MY DEAR SIR EVERARD,—I was overturned two days ago by a drunken coachman, but received no hurt. Lady Banks and my sister and I were driving home from dining with Sir A. Macdonald. We are all three rather heavy, and I, as you know, quite helpless. We were obliged to lie very uneasily at the bottom of the coach for half an hour before assistance could be got to lift us out. We all bore our misfortune without any repining or any demonstration of the follies of fear ; and we are all now quite recovered from the effects of our accident, except my sister, who has a cut on her head filled with lint and doing very well. But both the ladies have gone everywhere since, without an hour's confinement."

This accident was doubtless of more consequence than at first appeared. Mistress Banks died on September 27, after a " slight indisposition," which had not prevented her calling, a few days previously, upon the Princesses at

MISTRESS SARAH SOPHIA BANKS
From a drawing by John Russell, R.A.

Kew. At the age of seventy-four, she could not afford to be shaken about as described.

Mistress Sarah Sophia Banks was a zealous student of Natural History, and a most indefatigable collector of curiosities of all sorts. But her moral worth (says a cotemporary obituary), even more than her talents and knowledge, rendered her the object of esteem and regard to all who had the pleasure of being acquainted with her. The life-long attachment of Sir Joseph and his sister speaks well for both of them. And there seems to be no recognizable limit to the versatility of either one. Living almost always a member of his household, Mistress Banks was a devoted companion of Sir Joseph, proud of his ever-growing fame, and constantly at his service as amanuensis or interviewer. The amount of manuscript matter in her handwriting which has been preserved is really prodigious. One item alone in the British Museum Catalogue, which appears to be her handiwork, consists of no less than sixty-five volumes. These include, among other matters, papers on ceremonials, heraldry, and collections for a history of the Order of the Garter.[1] Another is a list of the books in Banks's library [2] (apparently that at Revesby); one section of which is a copious bibliography of archery. She also collected Broadsides,[3] books on Chess, Engravings, and news-cuttings galore. Her greatest triumph in this way was an extraordinary accumulation of Visiting Cards. This collection [4] is a perfect charm, as not only reflecting Miss Banks's cultured taste, but suggesting an elegant testimony to one of the many refinements of her period.

There are Visiting Cards, Invitation Cards, and

[1] Addl. MSS., 6277 to 6342.
[2] Addl. MSS., 33494.
[3] Nine volumes of Broadsides and Caricatures preserved in the General Library.
[4] Now kept in the Print Room of the British Museum. They must be at least ten thousand in number.

" Thanks for kind Inquiries." All branches of the Fine Arts are here represented, by engravings generally in the best possible execution. Nearly every object in Nature and in imaginative Art has been laid under contribution to furnish dainty designs. Vases, trees, gondolas, cupids, ruins, temples, ships, vie with birds, snakes, wild animals, agricultural and architectural objects ; set in scrolls, squares, ovals, wreaths, or festoons ; some embossed, many coloured ; some humorous, as if engraved for the purpose of being " collected " (including one of Sir Joseph Banks upon a tiny map of Iceland). Every country appears to be represented, France, Italy, and Germany contributing most largely to the store.

These worthy people lived well in the fashion. In her younger days Miss Banks went about in the best style. She dressed well, and drove a tandem in Hyde Park. But, according to one informant,[1] she lost the inclination for a specially elegant appearance out of doors. " She was looked after by the eye of astonishment wherever she went. Her dress was that of the old school ; her Barcelona quilled petticoat had a hole on either side for the convenience of rummaging two immense pockets, stuffed with books . . . both she and Lady Banks, in compliment to Sir Joseph, had their riding-habits of wool made out of his own produce ; in which dresses the ladies appeared on all occasions." A tall servant, with a taller stick in his hand, went with her everywhere. The same writer tells of his once taking an immense number of tradesmen's tokens to Soho Square ; with a note begging Mistress Banks to accept of any she might want. She politely called upon Mr. Smith, and thanked him ; but, strange as it might seem, she regretted to say that

[1] J. T. Smith : *A Book for a Rainy Day.*

J. Gillray

AN OLD MAID ON A JOURNEY

out of so many hundreds there was not one that she was in want of.[1]

Sir Joseph sent all his sister's coins, books, and curiosities to the British Museum, offering the selection of anything the Trustees did not already possess. As most of her collections were unique in their way, the Museum profited considerably by the opportunity.

The End was now in view. References to Banks occur in private letters between friends, bearing solicitude on account of his pains and his growing weakness, mixed with a sense of satisfaction that his strong and still vigorous mind supported him through it all. And, probably with sincerity, it was not uncommon to express an opinion of the great loss to Science death would prove.[2]

When the year 1820 opened, Sir Joseph was still at his post. He occupied the chair at the Council of the Royal Society on January 13, 20, and 27, and once in February. March 16 was the last day of his appearing in the seat which he had occupied for so many years with such distinction. On May 18 Sir Everard Home informed the Council that the President had sent in his resignation of the post. But it was resolved unanimously by those present (Goodenough, Bishop of Carlisle, in the chair), " That instead of accepting the resignation of the President, the Council do with one voice express their most

---

[1] Miss Banks had the distinction, common to all persons who were in any way " quaint " or " unconventional," of being numbered among the inspirations of James Gillray. The Caricaturist drew a humorous picture of *An Old Maid on a Journey* carrying a lap-dog and fan, and parasol ; servants follow, bearing packages, a bird-cage, a cat, etc. (published November, 1804). A reproduction of this appears on the adjoining page.

[2] One odd relic found among his papers (N.H. Mus.) is an anonymous letter to Banks, signed F.R.S. (March, 1819). After it is written with some pathos " From an old man to an old man." After pious allusion to the impending act of Providence in removing him, " We must look forward to that hour with calm fortitude," and begs him to nominate his successor !

cordial wishes that the President should not withdraw from the Chair of the Society, which he has filled so ably and so honourably during a period of forty-two years."

This message, doubtless, gratified the aged President. Willing to die in harness, and rejoicing at this last friendly touch on the part of his colleagues, he sent them a reply in the following terms :

"32 SOHO SQUARE, *June* 1

" Sir Joseph Banks begs leave to inform the Council of the Royal Society that his motive for offering his resignation of the office of President was a conviction that old age had so far impaired his sight and his hearing as to render him by no means so well able to perform the duties of that respectable office as he has been. He is gratified in the extreme by finding that the Council think it possible for him to continue his services without detriment to the interests of the Society, and he begs leave to withdraw his resignation, assuring the Council that his utmost exertions shall never be wanting to conduct, so far as may be in his power, the affairs of the Society."

On June 19, 1820, Barrow sent word to Marsden that " Sir Joseph breathed his last at seven o'clock." Thus ended a career of almost unexampled usefulness ; one that left the country better and wiser ; and the whole world indebted for steps in progress and culture to the life and labours of a most worthy English gentleman.

Records of Banks's personality are somewhat rarer than is usual with men who have enjoyed the public favour for a long series of years. The few casual references to him, that have appeared in print, leave him little more than a shadow as to the working of his mind or the inclinations of his heart. His numerous correspondents, almost without exception, display a regard for him that may be called really affectionate. They loved him, and rejoiced in his friendship ; but rare indeed is the exchange of deep sentiment. His remaining something of a phantom to us lies in this : that he never talked or wrote about himself. His letters, journals, instructions, scientific reasonings, are all objective. The first-person-singular is rarely used. When he has to make a protest, or a mild suggestion of blame, the error or the offence is not shown to be a wrong done to himself. It is made to appear a breach of proper principle, neither more nor less. And when anything bordered on misconduct which could not be condoned, it was Banks's way to dismiss from his thoughts the offence and the offender alike.

If, however, we cannot easily recognize the inner man, in the absence of self-revelation through the medium of diaries or letters, it is not difficult to understand why Sir Joseph Banks secured such general respect and affection. His life, as it appeared on the surface, was one long career of honour ; shown in his habitual honesty of purpose, his generosity and kindliness, and in his untiring public spirit. In one of the few cases where the above-mentioned law (as to the use of the first-person-singular) does not apply, we find him writing to Brissot de Warville : " My disposition is not of a nature inclined to wish for the control of any man's sentiments ; we should all speak or print what we think is true. For we shall certainly be esteemed by the world in proportion to the truth and justice of what we speak or print." These are golden words, and it was because of Banks's uniformly and instinctively living up to their tenour, that he kept such firm hold upon the regards of his fellow-countrymen. The circumstance that he did not please everybody can be accounted for, as long as the envy of distinction, or the jealousy of high personal qualities, continues to infect the meaner portion of humanity.

Banks had the advantage of a fine presence, tall and

well-proportioned. The best idea that can be gained of
him is through the portrait by Lawrence, to be seen in the
Trustees' Room at the British Museum.[1] " In his earlier
days, Sir Joseph exhibited a manly form, with a countenance that betokened intelligence, and an eye that
gleamed with kindness. His manners were courteous,
and his conversation was replete with instruction." [2]

Suttor, the botanical collector, who published at
Paramatta a short notice of Sir Joseph, describes him
as " always affable and kind ; of a very pleasing countenance, highly intellectual, a manly deportment, but very
little of the fine gentleman or the courtier ; a kind
master ; and a steady friend to high and low, rich and
poor . . . a cheerful and buoyant disposition, with
presence of mind and courage in distress." Alluding to
the doggerel of Peter Pindar, who could not possibly
spare so prominent a personage as Banks, Suttor remarks : " like his gracious Sovereign, he laughed at the
witty though virulent poet, and never caught a butterfly
the less."

Banks naturally became the object of numerous caricatures. But, as Cuvier says in allusion to this matter :
" Le seul remède applicable à de pareilles piqûres était
d'en rire. Ce fut celui qu'il employa." When Captain
Marryat was proposed to the Royal Society, some one
feared that his skits on Banks would be unfavourable to
his candidature. This was foolish. And coming to
Banks's ears he let it be understood that he admired
the drawings and purposely kept them on his table. " I
wouldn't be without them for the world," he said.[3]

One more " Impression " may be given, written by
one of his scientific friends, in the *Philosophical Magazine* :
" Sir Joseph in person was tall and manly, and his counte-

[1] *v.* Frontispiece.
[2] Faulkner : *History of Chelsea*, II, 190. This writer mentions a
portrait by Garrard, which has " an extraordinary degree of fidelity."
[3] Marryat's *Life*, by his daughter, I, 81.

nance expressive of dignity and intelligence. His manners were polite and urbane ; his conversation was rich in instructive information, frank, engaging, unaffected, and without levity, yet endowed with sufficient vivacity. His information was general and extensive. On most subjects he exercised the discriminating and inventive powers of an original and vigorous mind ; his knowledge was not that of facts merely, or of technical terms, and complex abstraction alone, but of science in its elementary principles, and of nature in her happiest forms."

In France, Belgium, Sweden, Germany, Italy, Russia, Switzerland, everywhere that the merits of a good and great Englishman could be eulogized, it was done. Science all over the world had lost its Nestor, and she stinted not the language of universal goodwill toward the memory of one who had drawn the nations together, in a field apart from the sordid struggles of life. His own supremacy was of that character which does not lay its rivals in the dust ; his path through life was strewn with the trophies and triumphs of benevolence. Something like this was the universal estimate of Europe.

One of the more eloquent of these testimonies to the worth of Sir Joseph Banks is that of Jean Baptiste Biot, the famous physicist. He spent much of his scientific life, in his younger days, in England ; and died in 1862, eighty-eight years of age. This is his memorial of his old friend : [1]

" Que ne puis-je peindre ce que je sentis en voyant pour la première fois ce vénérable compagnon de Cook ! Illustre par de longs voyages, remarquable par une étendue d'esprit et par une élévation de sentimens qui le font s'interesser également aux progrès de toutes les connaissances humaines, possesseur d'un rang élevé, d'une grande fortune, d'une considération universelle, Sir

[1] *Mélanges Scientifiques et Littéraires,* I 77.

Joseph a fait de tous ces avantages la patrimonie des savans de toutes les nations. Si simple, si facile dans sa bienveillance, qu'elle semble presque, pour celui qui l'éprouve, l'effet d'un droit naturellement acquis ; et en même temps si bon, qu'il vous laisse tout le plaisir, toute l'individualité de la reconnaissance. Noble exemple d'un protectorat, dont toute l'autorité est fondée sur l'estime, l'attachement, le respect, le confiance libre et volontaire ; dont les titres consistent uniquement dans une bonne volonté inépuisable et dans le souvenir des services rendus ; et dont la possession longue et non contestée fait supposer de rares vertus et une exquisse délicatesse, quand on songe que tout ce pouvoir doit se former, se maintenir, et s'exercer parmi des égaux."

Sir Joseph was buried unostentatiously, at his own request. He died at Spring Grove, in Heston parish ; in the churchyard of which his mortal remains were laid. By his will, Robert Brown was provided for by annuity ; and was given the use of Banks's library and Herbarium, with reversion to the British Museum. Frank Bauer,[1] the clever painter of plant-life, who was so long associated with Banks, likewise received an annuity by bequest.

Sir Joseph's portrait was painted several times. Two early pictures are by West and Reynolds. The former is a full-length figure, cloaked in a South Sea island robe ; the rendering is dignified and animated, while the surroundings are evidently meant to be a reminder of his character as a traveller. Sir Joshua's picture must be

[1] Frank Bauer painted flowers with a marvellous gift. His drawings of botanical physiology surpassed those of any who had previously painted the vegetable world. He took up his residence at Kew about 1789, with the patronage and encouragement of the King and of Sir Joseph Banks. The Royal Princesses were devoted to Bauer, and coloured some of his outlines of rare flowers. So his title of Botanical Painter to His Majesty was no sinecure. His brother Ferdinand, who accompanied Captain Flinders, had also a very superior talent of the same sort.

of nearly the same date. Banks is sitting at a table, surrounded by geographical indications. A fine mezzo-tint by Dickinson was the means of popularizing this portrait. Another portrait by Reynolds is one of the group of seven members of the Dilettanti Society.

George Dance painted Banks as in later middle life ; as also did Garrard, whose picture is said to be a very good likeness. Sir Thomas Lawrence made a crayon drawing of Banks, now in the National Portrait Gallery. Thomas Phillips produced several portraits. One of these is the property of the Royal Society, another hangs in the National Portrait Gallery, a third in the Boston Town Hall, and one (painted in the last year of Sir Joseph's life) belongs to the Royal Horticultural Society. John Russell made a crayon drawing, which was engraved more than once.

The marble statue by Chantrey, which stood for many years in the Entrance Hall of the British Museum, has been removed to the Natural History Department in Cromwell Road. In the opinion of Sir Henry Ellis, this statue offered by far the best picture of Banks. There are busts in the Natural History Museum, in the laboratory at Chelsea Gardens, and in the Banksian Library at Bloomsbury. This last is by Mrs. Damer, presented by her to the Museum in 1814. There is also a medallion portrait by Pistrucci, chief engraver at the Mint (Egerton MSS., 2851).

THE END

INDEX

INDEX

A

Abel, Dr. C., naturalist, 277, 298
Abercorn, Lord, 262
Aberdeen, 96
Abernethy, Dr, 287
Abyssinia, 174, 283, 284
Académie Française, 82, 84
Adams, Thomas, Captain, R.N., 10
Aerostation, 71
Africa, 142, 145–53
African Association, 145–52, 302
Afzelius, Adam, botanist, 153–5
Agriculture : *Distemper among horned Cattle*, 4
Agriculture, as a branch of botanical science, 12. *v.* "Farming"
Aiton, William, at Kew Gardens, 93
Aiton, W. T., 209, 260, 277
D'Alembert, Jean, quoted, 73
Aleppo, 116
Allen, Dr., gout specialist, 166
Allen, John, 230, 235
Almack's, 176 n
America, 140, 141, 211, 214
American Declaration of Independence, 81, 82
Amherst, Lord, 271, 277
Amiens, truce of, 149, 208
Anderson, Dr. Alexander, botanist, 121, 122
Anderson, Alexander, traveller, 151, 152
Anderson, William, surgeon, 40, 49
Angling, in Lincolnshire, 307
Anguish, T., 76
Antiquities the handmaid of history, 67
Apothecaries' Garden, Chelsea, 6, 93

Apple (Spring Grove Codling), 321
Arabic language, 150, 152
Arabs, 147
Arbitration, 263–4
Archæology, 3, 13, 314, 315
Archery at Revesby, 307
Arcot, India, 114
Arenberg, Duke of, 61
Ashby, Rev. George, 21, 53, 62
Astronomers, on the transit of Venus, 14
Astronomical Society, 258
Astronomy at Court, 63, 64
Aubert, Alexander, 56, 71, 78, 91
Auckland, Lord, 136, 207
Austin Friars, 56
Australia, 30, 217 *et seq.*, 226, 229
Ayrshire, 114
Azores, 297

B

Babbage, Charles, 175, 256
Bacstrom, Sigismund, M.D., 26 n, 167, 168
Baily, F. H., 256
Ballooning, 71
Bankes, of Kingston Lacy, 13
Banks, Joseph (d. 1727), notice of him, 3, 4
Banks, Joseph (d. 1736), 4
Banks, Sir Joseph, b. 1743, 3 ; left Oxford University, 1763, 8 ; Society of Arts, 1764, 121 ; voyage to Newfoundland, 1766, 11 ; F.R.S. 1766, 10 ; circumnavigated the world with Captain Cook, 1768, 14 *et seq.* ; voyage to Iceland, 1772, 32 ; Soho Square, 1777, 62 ; President of the Royal Society,

1778, 57; marriage, 1779, 62;
baronetcy, 1781, 63; Knight
of the Bath, 1795, 181; Privy
Councillor, 1797, 159; French
Institut, 1801, 210; Royal
Horticultural Society, 1804,
260; death, 1820, 326
Banks, Marianne (born Bate),
notice of her, 4; residence at
Chelsea, 6
Banks, Lady (b. Hugessen), 62,
64, 70, 99, 158, 258 n, 271, 272,
293, 317, 318, 324
Banks, Sarah Sophia, 63, 70, 71,
99, 258 n, 293, 307, 317, 322–5
Banks, William, 4, 6
Barbados, 124, 125, 138
Bardney, Lincs, 3
Barker, Rev. Thomas, 316 n
Barlings, Lincs, 3
Barra, W. Africa, 147
Barrington, Hon. Daines, 27, 37, 40
Barrow, Sir John, 152, 238, 258,
278, 281–2, 287, 291–4, 301, 318
Barrows, near Revesby, 315
Barton, G. B., 213
Bass, George, 230, 233
Batavia, 17–21, 40 n, 130, 131,
134, 278
Batavian Society, Rotterdam, 36
" Bath Butterfly," 180
Bath, 13
Bath, Lord, 99
Bathurst, Australia, 229
Bathurst, Lord, 106, 107, 296
Baudin,Captain, circumnavigator,
233, 236
Bauer, Ferdinand, 228, 230, 234,
238–41, 330
Bauer, Frank, 330
Bavaria, 259
Bayly, Wm., astronomer, 27, 40
Beaufort, Duke of, 99
Beaufoy, Henry, 145
Beaumont, Sir George, 38
Beckett, Thomas, bookseller, 28
Beddoes, Thomas, M.D., 188–191,
260
Beddoes, T. L., 188
Beechey, Capt. Frederick, 152
Behar, India, 113
Belles Lettres Society, proposed,
262
Belzoni, traveller, 284, 285, 287

Bengal, 113, 142
Benin, W. Africa, 278
Berkeley, Lord, 99
Berlin Academy, 84
Berne, 53
Berry, Miss, quoted, 173
Bessestedr, Iceland, 33
Biblical botany, 162
Bilbao, 102, 103
Billings, Joseph, 146
Biot, Jean B., 329
Blagden, Dr. Charles, Kt., 37,
64, 69, 71, 74, 77, 78, 79, 258 n,
287, 291
Blanchard, Jean P., his Aeros-
tatic Academy, 71
Bligh, William, admiral, 40, 49,
126 *et seq.*, 222–3, 226–7, 280–1
Bligh, Mrs., 223
Blosset, Miss, 22
Blumenbach, Prof. Johann F., 152
Board of Agriculture, 100–1, 319
Board of Longitude, 24 n, 159
Board of Trade and Plantations,
159, 220, 221
Bonaparte, Napoleon, 205
Boston, Lincs, 312, 313
Boswell, James, 37, 160
Botanical collectors, 95, 96, 139,
248, 274–7
Botany, at Oxford and Cam-
bridge, 6–8; South Sea Islands,
17, 18; Iceland, 33; Wales, 37;
Tranquebar, 45; Yorkshire,
47; at Court, 83; Soho Square,
90; British Colonies, 96; India,
114 *et seq.*; New South Wales,
226 *et seq.*
Botany Bay, 167, 192, 213, 214 n,
216, 230, 233
Botley, Hants, 107
Boucher, R., geographer, 53
de Bougainville, L. A., 28 n
de Bouillé, Marquis, 121, 122
Boulton, Matthew, engineer, 159,
278
Bowie, James, botanist, 276–7
Brazil, 277 n
Bread-fruit tree, 19, 123, 124, 127,
133, 137, 138
Bretschneider, Emil, quoted, 282
Bridgewater, Somerset, 13
Briscoe, Peter, 16
Brissot de Warville, Jean P., 327

Bristol, 13, 188, 191
Bristol Naturalists' Society, 12 n
Bristol, Lord, 45, 99
Britannia antiqua, 246
British Jungermanniæ, The, 298
British Museum, 30, 53, 59, 62, 65, 66, 67, 71, 74, 194, 209, 243, 254, 274, 284, 286, 309, 314, 323, 325, 328, 330, 331
Britten, James, botanist, 30
Broadley, A. M., 102 n
Broadsides, Miss Banks's collection of, 71, 323
Brocklesby, Dr. Richard, 77
Brodie, Sir Benj. C., 256, 287, 289
Brompton Botanic Gardens, 225
Brooke, Robert, 112, 113
Brougham, Henry, Lord, quoted, 5, 9, 24, 73, 75, 207, 211 ; notice of him, 302-3
Brougham, Henry, senr., schoolfellow of Banks, 5
Brown, Robert, botanist, 25 n, 228, 230-4, 238-43, 288, 300, 302
Brown, William, mutineer, 134
Bruce, James, traveller, 39, 174
Brussels, 61
Bryant, Jacob, antiquary, 66, 67
Buchan, Alex., artist, 16, 19, 30
Buckland, Dr. William, geologist, 302
Buitenzorg, Batavia, 278
Bulstrode House, Bucks, 65, 67
Burckhardt, John L., 152
Burgis, Thomas, artist, 34
Burke, Edmund, 160
Burney, Dr. Charles, 23
Burney, Fanny, quoted, 23, 41, 64, 67
Burney, James, R.N., 23, 40, 49, 51
Bute, Earl of, as horticulturist, 93

C

Cairo, 147, 284, 285
Calais, 71, 103, 104
Calcutta, 113, 115, 159, 273, 274
Caley, George, botanist, 224-9, 238, 276, 282
Cambridge, 8, 43
Camelford, Lord, 144
Canada, 98, 214, 318
Canton, 266, 269, 272, 273

Caoutchouc, early mention of, 16
Cape of Good Hope, 26, 45, 49, 96, 97, 140, 142, 219, 231, 242, 269, 277, 292
Cape Horn, 141, 144
Caricatures, 175 et seq., 325
Carlisle, Sir Anthony, 287
Carnatic, 114, 116
Cartwright, George, Labrador trapper, 35
Cartwright, John, R.N., 35
Caserta, 62, 198, 199
Cassel, Germany, 54, 169, 170, 171
Castile, 102
Castlereagh, Lord, 248
Cavendish, Hon. Henry, 56, 76, 78, 91, 210, 287
Cayenne, 208
Ceylon, 268, 273
Chambers, Sir W., architect, 93
Chantrey, Sir F., 331
Charing Cross, 165
Charles II, F.R.S., 59
Charretié, 207
Chatham, 140, 253
Chatham, John, 2nd Earl, 137
Chelsea, 6, 93, 186, 331
Cheltenham, 98, 179
Chemical Society, 255
Chemistry applied to medical science, 188-90
Chepstow, Gloucester, 13
Chester, 37
Chestnut tree management, 321 n
Chettle, Dorset, 13
Children, J. G., quoted, 65
Chile, 145, 230
China, 140, 235, 266 et seq., 298, 318
China ware, 272
Christ Church College, Oxford, 302
Christchurch, Hants, 93
Christian, Edward, 135
Christian, Fletcher, 134, 135, 136
Church Missionary Society, 142
Circumnavigation voyages: popularity of, 14
Cirencester, 98
Clarke, Dr. Thomas, 122
Clergy, eighteenth-century, 38
Clerke, Lieut. Charles, 24, 25, 29, 49, 50
Clevedon, Somerset, 13
Cleveland Hills, Yorkshire, 46
Cleveley, John, artist, 26 n

z

Clifton, Gloucester, 13
Club, the, 70, 90
Cobbett, William, 107, 211, 212
Coke, Thomas, 109
Coke, Lady Mary, quoted, 22 *n*
Cole, Rev. W., 174
Colman, George, 46, 47, 215
Colman, George, the younger, 46–8
Congo River, 278
Connecticut, 53, 146
Constantinople, 71, 216
Conybeare, Dr. William, geologist, 302
Cook, Elizabeth, 50, 52
Cook, James (father of Captain Cook), 48
Cook, Captain James, 15–20, 23, 27, 28, 29, 30, 31, 39, 40, 45, 48, 49, 50, 51, 53, 61, 80 *n*, 110, 129, 130, 143, 167, 169, 176, 177, 185, 213, 234, 329
Cooper, Gloucester clothier, 99
Cooper, Thomas, 85, 88
Copenhagen, 66, 222, 245, 247
Corneille, Daniel, 111, 112
Cork, 247
Cornwallis, Charles, Earl, 113
Coromandel, xi, 115
Correa de Serra, Josef, 231
Cotswold Hills, 99
Country fair at Revesby, 306
Country parsons, eighteenth-century, 38
County history, importance of, 4
Coupang, 131
Coxe, William, traveller, 146
Cradock, Joseph, quoted, 23 *n*, 43 *n*
Cranberry cultivation, 261, 317
Cranch, —, naturalist, 279
Crichel, Dorset, 11
Criticism, inept, 18
Croker, J. W., 37, 238
Crosley, John, astronomer, 231–2
Crown and Sceptre, Strand, 70
Cullum, Sir John, 38, 39, 40, 43, 57
Cunningham, Allan, botanist, 276, 277
Curtis, Wm., xi
Cust, Lionel, quoted, 38
Cuvier, Baron, quoted, 49, 328

D

Dalrymple, Alexander, geographer, 15, 78, 214, 291
Damer, Mrs., 331
Dampier, Captain William, 18
Dance, George, painter, 331
Danish East Indies, 114, 134
D'Arblay [=Burney, F.]
Dartmouth, Earl of, 261
Darwin, Charles, 319
Davy, Sir Humphry, 256, 260, 265, 293, 300, 301
Davy, Dr. John, 301
Delany, Mrs., quoted, 66; her plants in paper mosaic, 67
De Candolle, 299, 300
De Caen, Governor of Mauritius, 236, 237
Deer-park economy, 308, 320 *n*
Denmark, 244–6, 252
Denne, Dr., 102
Deptford, 129, 140
Dettingen, 55
Devonshire, Georgiana, Duchess of, 189, 190
Dickson, John, horticulturist, 260
Dilettanti Society, 38, 158, 319
Dillwyn, John, 105, 108
Dog story, 62
Dolomieu, D. Gratet de, naturalist, 197, 205–7
Dominica, W. Indies, 120
Dorrington, Lincs, 313
Dorsetshire, tour in, 12
Douglas, Dr. John, 51, 54, 184
Dover, 103, 104
Downing, Flintshire, 9, 60
Downton, Hereford, 319
Dryander, Jonas, xi, 65, 71, 97 *n*, 181, 232, 242, 243, 288
Dublin, 112
Ducie, Lord, 99
Duckett, William, farmer, 100

E

East, Hinton, horticulturist, 122–4, 137
East India Company, 111–6, 157, 159, 214 *n*, 269, 271, 273, 274, 277, 290
Edgar, Thomas, 54

Edgeworth, R. L., 188, 191
Edinburgh, 34, 55, 139, 145, 149, 162, 231, 247
Edinburgh Review, 302, 303
Edwards, Captain, 135
Edwards, Bryan, 122
Egypt, 283, 284
Egyptian antiquities, 284–6
Ehret, George D., painter of plant life, 66
Elgin, Lord, 209
Ellis, John, botanist, 66
Ellis, Sir Henry, 331
Ellis, W., surgeon, 52
Endeavour Straits, 128, 132
Engineers' Society, 158, 160
Englefield, Sir Henry, 38, 91
D'Entrecasteaux, Joseph, circumnavigator, 192
Esher, 100 *n*
Esquimaux, 34, 36, 244
Etches, R. C., 140
Ethan, the wisest of men, 157
Ethnographical speculations, 36
Eton School, 5
Evelyn, John, 39, 98

F

Fabricius, Jean C., naturalist, 65–6
Falconer, Sir Thomas, 16
Falkland Islands, 21
" Farmer George," 98
Farming, 99–101, 110, 308 *et seq.*
Faroe Islands, 248
Faujas de Saint-Fond, Barthélemy, 89–92
Fawkener, 220
Felons, transportation of, 213–6
Fen-land reclamation, 308, 311
Ferdinand, King of Naples, 62, 198, 199
Fezzan, 147, 152
Fielding, Sir John, 7
Fifeshire, 231
Fishing-parties at Revesby, 307–8
Flinders, Captain Matthew, 230 *et seq.*, 278, 280, 313
Flora Anglica (Wm. Hudson), 13
Foreign travels and visits in England, 63, 65, 89, 90, 171, 299
Forrest, Captain Thomas, 290
Forster, Georg, 27, 54, 169–71

Forster, John Reinhold, 27, 34 *n*, 40, 169, 170
Forsyth, William, botanist, 260
Fothergill, Dr. John, 29, 66, 139, 183, 184
Fowler, William, architect, 313–4
Fox, Rt. Hon. C. J., 215
Franklin, Sir John, 234, 235, 313
Frederick the Great, 169
Frederick, Princess of Wales, 93–4
Freetown, West Africa, 153
French Directory, 195
French oversea discoveries, 142, 233, 236
French relations with England, 207, 210, 237
French Revolution, 86, 188, 194, 195, 244
de Fréville, A. F. J., 28 *n*
Friendly Islands, 19, 128
Fryer, John, 130
Furneaux, Captain Tobias, 40, 41

G

Galt, John, 283
Gambia River, 147, 148, 151
Garrard, George, painter, 328, 331
Geikie, Sir Archibald, 89
Geographical Club, 153
Geological Society, 255, 257, 302
George III, King, 21, 22, 25, 47, 51, 55, 64, 67, 79, 94, 96, 98–103, 105, 108–10, 124, 136, 137, 173, 175, 178–9, 221, 301, 311, 312
George, Prince Regent, 174, 265
Georgium Sidus (Uranus), discovery of, 63
Gerarde's *Herbal*, 6
German appreciation of Captain Cook, 30, 54
Gillray, James, 180, 325
Gilpin, George, 81
Glasgow, 298
Gloucestershire, festivities in, 98–9
Goethe, von, W., 242
Good, Peter, gardener, 230, 232–3
Goodenough, Dr., Bishop of Carlisle, 325
Gore, Lieutenant John, 51
Goree, W. Africa, 151
Gossett, I., 77

Göttingen, 53
Gough, Richard, antiquary, 57, 314
Graefer, John, horticulturist, 199, 204
Grampound, Cornwall, 4
Graufell, Iceland, 33
Gravesend, Kent, 24
Gray, Asa, botanist, 299
Green, Charles, astronomer, accompanied Captain Cook on his first voyage, 15
Green, John, Bishop of Lincoln, 43
Greenland, 248
Greenwich, 45
Gregory, Dr. Olinthus, 256
Grenada, W. Indies, 120
Greville, Hon. C., 36, 38, 66, 184, 232, 260, 319
Greville, Henry Fulke, 262
Greville, R. F., 108–10
"Grey Wethers," near Marlborough, 13
Grimsby, Lincs, 3
Grose, Major, 219
Guines, France, 71
Gunning, J., 77

H

Hague, The, 137
Hackman, James, 43
Halesworth, Suffolk, 253, 254, 294, 299
Halifax, Nova Scotia, 139
Hall, Captain Basil, 152
Halle, Germany, 169, 170
Hallett, John, 136
Ham, Surrey, 100 n
Hamilton, Emma, Lady, 197, 200–5
Hamilton, George, 135 n
Hamilton, W., orientalist, 283–6
Hamilton, Sir William, 38, 61, 197–206, 307, 319
d'Harcourt, Duke, 193, 194
Hardcastle, Rev. Joseph, 251
Hardwick, Suffolk, 38
Harlock, —, 45
Harrison, George, 275
Harrow School, 5
Hastings, 208
Hastings, Warren, 113
Hartford, U.S.A., 53

Hatfield, Herts, 304
Havnefiord, Iceland, 33
Hawkesbury, Lord, 247
Hawkesbury River, N.S. Wales, 219
Hawkesworth, Dr. John, Editor of Captain Cook's first voyage, 30, 31
Hawkey, Lieutenant John, 279
Hawsted, Suffolk, 38
Heberden, Dr., 79
Hebrew traditions of wise men, 157
Heidelberg, 66
Hekla, Mount, ascended by Banks, 33–4
Herbert of Cherbury, Lord, 38
Herculaneum MSS., 265
Herschel, Caroline, 91
Herschel, Sir John, 256
Herschel, Sir William, 63, 64, 90, 210 287, 291
Hessian fly, 320 n
Heston, Middlesex, 316, 330
Highbury, Middlesex, 56
Hinchingbrook, Hants, 23
Hobart, Lord, 150
Hodges, William, 27
Hogan, J. F., 254 n
Hollis, Brand, 77
Home, Sir Everard, 5, 280, 281, 287, 304, 322, 325
Hooker, Sir J. D., xi, 17, 113, 299
Hooker, Sir Wm. J., 34, 248–53, 282, 294–9
Hooper, James, 277
Hope, Dr. John, Edinburgh professor of Botany, 83, 114, 118, 122, 139, 140, 257
Horncastle, 312, 313
Hornemann, Frederik, 148–9, 216
Hornsby, Rev. Thomas, Savile astronomer, 14, 63
Horsfield, Dr. Thomas, botanist, 273, 274, 282, 296
Horsley, Dr. Samuel, 75, 76, 77, 78, 80, 81, 179, 211, 212
Horticulture, 93, 94, 95, 112, 120 et seq., 138, 199, 260, 261, 308, 319, 320
Hortus Kewensis, 83, 277
Houghton, Major Daniel, 147, 148
Hounslow, Middlesex, 7
Houston, William, xi

Howick, Lord (Earl Grey), 208
Hudson, William, botanist, 13, 225
Hudson's Bay, transit of Venus observed there, 15
Hudson's Bay Company, 35
Hugessen, Wm. W, 62
Humboldt, von, Alexander, 242
Hume, David, 162
Humours of success in life, 172 *et seq.*
Hunt, —, 98
Hunter, Captain John, 218, 222
Hunter, Dr. Alexander, 39, 314
Hunter, John, 66
Hussey, Rev. T. J., 175
Husson : *eau médicinale*, 281
Hutton, Dr. Charles, 74, 75, 77, 78, 79
Hutton, Dr. James, geologist, 36
Hyde Park skating, 117 ; driving, 324
Hyder Ali, 117

I

Iceland, 32 *et seq.*, 68, 181, 244, 294, 295, 324
Indian Serpents, 120 ; cement for mending noses, 158
Institut National, 207–8, 210, 212
Islip, Oxfordshire, 302
Italy, 197, 198

J

Jamaica, 96, 122, 123, 124, 136, 137, 138, 177 *n*
Japan, 140
Jardin des Plantes, 207
Java, 20, 129, 192, 273, 274, 296
Jeffries, J., American balloonist, 71
" Jemmy Twitcher," 23
Jodrell, P., 77
Joe Miller, quoted, 297
Johnson, Henry, inventor of Logography, 165, 166
Johnson, Dr. Samuel, 37, 41, 65, 160
Johnston, Major George, 223
Johnstone, Cochrane, 106, 107
Jones, Dr. H. Bence, 258 *n*
Jörgensen, J., 230 *n*, 251–4
Jussieu, Antoine L., 82, 197

K

Kæmpfer, Ingelbert, xi
Kaffir country, 97
Kamtchatka, 50, 146
Kangooroo, the, 18
Kaye, Richard, Dean of Lincoln, 10, 12
Kendall, Rev. Thomas, 142
Kerr, William, botanist, 266, 268, 269, 282, 302, 318
Kew Gardens, 92, 93–6, 98, 102, 145, 154, 209, 217, 225, 229, 235, 266, 274, 275, 297–9, 319
Kew Herbarium, 227, 296, 298, 299
Kew Palace, 21, 103, 105
King, Captain Philip G., 218, 228, 234, 238
King, Captain James, 51, 184
King, Philip Parker, 277 *n*
Kingston, Jamaica, 122
Kingston, St. Vincent, W.I., 121
Kingston Lacy, Dorset, 13
Kippis, Dr. Andrew, 80
Kirkleatham, Yorkshire, 48
Kirkstead, Lincs, 3
Kitson, Arthur, quoted, 24, 40
Knight, Richard Payne, 38, 173, 262, 319
Knight, Thomas Andrew, horticulturist, 260, 261, 282, 319, 320
König, John Gerard, botanist, 114, 115, 116, 118, 119
Königsberg, 146
Kyd, Colonel Robert, 113, 115

L

de La Billardière, J. J., 192–7, 212
Labrador, 10–12, 34, 35
Lamb, Charles, 269
Lambourn, Essex, 43 *n*
Lampooning in fashion, 175
Lanarkshire, 93
Lance, David, 266, 267, 268, 271
de La Pérouse, P. P., circumnavigator, 192, 216
Laplanders, 244
Latham, Dr. John, 283
Laugarvatn, Iceland, 33
Lauragais, B., 28

Law reform, 83
Lawrence, Sir Thomas, R.A., 331
Leach, —, 32
Lecky, W. E. H., historian, 217 n
Ledward, Thomas, 130, 134
Ledyard, John, 53, 146, 147
Lee, Arthur, 81, 82
Leslie, John, professor of mathematics, 162
Leith, 247, 248
Lever, Sir Ashton, 38
Lewisham, Kent, 56
Leyden, 55
Lhasa, 271
Lightfoot, Rev. John, botanist, 9, 10, 37, 39, 66
Lincoln, 3, 310, 314
Lincolnshire, 3, 9, 146, 159, 231 n, 304 et seq., 320 n
Lind, Dr. James, 27, 32, 34, 64, 67, 68
Linnæus, Carl, 16, 65, 67 n, 114, 117, 225, 257
Linnean Society, 7, 82, 97 n, 138, 155, 256, 257, 273
Lisbon, 11
Liverpool, 239, 241
Liverpool, Charles, first Earl, 159, 311 ; R. B., second Earl (= Hawkesbury), 247, 276
Livonia, 114
Lloyd, J., 212
Logography, 165
London Bridge, as seen by Esquimaux, 35
Lort, Rev. Michael, 53
Loten, J. G., 170
Louis XVI, 195, 216
Louis XVII, 193, 195
Loyalist Americans, 214, 215
Lucas, Simon, 147, 216
Lunardi, V., the balloonist, 71
Luton Hoo, Bedfordshire, 93
Luzon, 268
Lyon, Emma (=Hamilton)
Lyons, Israel, 8
Lysons, Dr. Samuel, 98
Lyttleton, Charles, Bishop of Carlisle, 10

M

Macaronis, 176
MacArthur, John, 219–23

Macartney, Lord, 292
Macaulay, T. B., 22 n
Macdonald, Sir A., 322
Mackenzie, Sir George Stewart, 250, 253 n
Maclean, —, of Drumnen, 32
Macquarie River, Australia, 277
Madeira, 17, 232
Madocks, William A., 107
Madras, 15, 114, 116, 119
Madrid, 103
de Magellan, J. H., 60, 61
Mahon, Lord, 180
Maiden, J. H., botanist, viii, 213
Malacca, 113–5
Manchester, 224
Mann,—, 61
Mann, Horace, 174
Manning, Thomas, 269, 270, 298
Marcet, Dr., 299
Marr (or Matra), J., his curiosities brought to Banks, 45
Marryat, Captain F., 152, 328
Marryat, Mrs. Joseph, 319
Marsden, Rev. Samuel, 221
Marsden, William, 152, 239, 241, 287, 290, 292, 326
Martin, Joseph, Fr. naturalist, 208
Markham, Admiral Sir C. R., 151, 270
Martyn, John, botanical professor at Cambridge, 8
Marylebone Church, 201
Maseres, Francis, 76
Maskelyne, Rev. Nevil, astronomer-royal, 14, 45, 63, 68, 76, 91, 210, 290
Masson, Francis, botanical collector, 96–8, 297
Matra, J. M., 40 n (45), 214–6
Maty, Paul Henry, 57, 73, 74, 78
Maty, Dr. Matthew, 73
Mauritius, 236, 237
Medals : Copley granted to Captain Cook, 49 ; to Sir B. C. Brodie, 287 ; Royal Society, Cook commemoration, 51 ; Board of Agriculture, 100 ; Society of Arts, 121
Medina, Africa, 147
Melville, General Robert, 120, 121
Menzies, Archibald, 139, 140, 143, 144, 145
Merino sheep, 104, 105, 106, 311

Merino Society, 312
Mesmerism, 83
Metcalfe, Philip, 160
Midlothian, 116
Miller, John Frederick, artist, 26, 32, 34, 181 *et seq.*
Miller, James, draughtsman, 26 *n*
Mile End nursery garden, 199
Miller, Philip, horticulturist, 93, 199
Mitre tavern, 56
Mola, Iceland, 33
Monboddo, Lord, his ethnographical studies, 36
Mogador, 152
Monastic houses in Lincolnshire, 3
Monkhouse, William B., surgeon, 20
Monkhouse, Jonathan, 20
Monson, Lord, 307
Montagu, Edward Wortley, 22
More, Hannah, quoted, 37
Moravian Brethren, 45
Morocco, 147, 152
Morrice, —, 22
Morris, Captain Thomas, 180
Morris, Valentine, 13, 121, 123
Morrison, Dr. Robert, 271
Morton, Charles, librarian, British Museum, 10
Moseberg, T. H., 168
Mount Edgcumbe, Cornwall, 11
Mouron, —, 104
Mulgrave, Lord [= Phipps]
Mulgrave Hall, Yorkshire, 46, 48
Mull Island, 32, 92
Murdoch, T., 282
Mutiny of the *Bounty*, 131 *et seq.*
Mutiny in N.S. Wales, 222

N

Naples, 62, 197
Natural history studies, 5, 9, 10, 11, 12, 18, 33, 34, 37, 58, 59, 65, 90, 116, 117, 119
Natural History Museum, 18, 30, 33, 34, 66, 277, 331
Nautical Almanac, 159
Naval Architecture, Society for Improvement of, 158
Needham, 61
Nelson, Lord, 204, 209, 222

Nelson, David, botanist, 126, 129, 130, 132, 134
New Burlington Street, 9, 12, 61
New Guinea, 128, 290
New Holland, 17, 128, 142, 149, 213, 214 *n*, 225, 230, 231, 233, 240, 241, 242, 251
New South Wales, 29, 136, 213, 215, 217–24, 226, 229, 241, 277
New York, 81
New Zealand, 17, 40, 128, 141, 142 *n*, 277
Newfoundland, 10, 12, 35
Newton, Sir Isaac, 175, 178
Nichols, John, 22 *n*, 57, 85
Niger River, 147, 150, 151, 278
Nootka Sound, 142, 144
Norfolk Island, 241
North Pole, expedition by C. Phipps, 8 ; Banks proposes a voyage, 36
North, Lord, 214, 215
North, Mrs. Brownlow, 185–7
North-West Passage, 49
Northampton, Marquess of, 287
Northumberland, Duke of, 187
Norton, Kent, 62
Norwich, 257, 299

O

Oberea, Queen, 177
Oldaker, Isaac, gardener, 318
Omai, an Otaheitan, in England, 41 *et seq.*, 46, 47
Opie, John, R.A., 178 *n*
Orange culture in Australia, 229
Otaheite, 17, 19, 20, 28, 40, 50, 128, 130, 133, 134, 136, 137, 141, 177
Oxford University, 7, 8, 188, 189
Oxley, John, explorer, 277

P

Pacific Ocean, 141
Paddington, 319
Pæony, the, 268, 318
Pallas, Dr. P. S., 146
Paradise Walk, Chelsea, 6
Paramatta River, 219, 226, 328
Paris, 82, 83, 90, 210
Paris, Dr. J. A., quoted, 301

Park, Mungo, 148–52, 216, 219, 231, 284
Parkinson, Sydney, artist, 16, 20, 21, 29, 30
Parkinson, Stanfield, 29
Parry, Captain W. E., 294, 301
Party spirit, 86, 87, 223
Patton, Colonel Robert, 113
Paul, Sir George, 99
Peake, R. B., quoted, 47, 48
Pearce, Nathaniel, 283, 284
Peebles, 149, 150
Pennant, Thomas, 9, 16, 21, 27, 33, 37, 60–6, 170, 181–5
Pepys, Sir Lucas, 190
Perthshire, 139
" Peter Pindar," 173, 177, 178
Peterborough, 4
Petersham, Surrey, 100 n
Phelps, Samuel, 251
Phillip, Captain Arthur, 216, 217
Phillips, Molesworth, R.N., 49
Phillips, Thomas, R.A., 313, 331
Phipps, Constantine (afterward Lord Mulgrave) ; his polar Expedition, 8 n ; to Newfoundland with Banks, 10, 11, 35, 43, 45, 215 ; Yorkshire tour with Banks, 46 et seq.
Phipps, Augustus, 46
Pisania, W. Africa, 151
Pistrucci, B., medallist, 331
Pitcairn Island, 135
Pitcairn, Dr., 56, 119, 139
Pitt, William, Right Hon., 206
Planta, Joseph, 56, 74, 291
Plants—
    Albiflora fragrans, 318
    Araucaria imbricata, 145
    Artocarpus incisa, 19
    Cassa acutifolia, 124
    Drosera longifolia, 225
    Ericaceæ, 97, 145
    Hedysarum gyrans, 92
    Heliconia, 183, 184
    Hydrangea hortensis, 318, 319
    Massonia, 183
    Menziesia ferruginea, 145
    Rosa banksiæ, 302, 318
    Sequoia sempervirens 145
    Solanum tuberosum, 320 n
    Vaccinium macrocarpum, 321
    Zizania aquatica, 318

Plants, English, acclimatized in Jamaica, 123
Plant collectors, 96, 97, 112 et seq., 126, 143, 153, 226, 243, 295
Plymouth, 11, 17, 32, 35, 40, 135
Pneumatic medicine, 188–91
de Poederle, Baron, 61
Porcelain, 272, 273
Port Jackson, 142, 216, 219, 232, 234, 235, 238, 240
Porteous, Henry, 112
Portland, Duchess of, her museum, etc., 65, 66
Portlock, Lieutenant N., 137
Portsmouth, 105, 135, 151
Portugal, 11, 97, 102, 103
Price, Major William, 193
Priestley, Dr. Joseph, 24 n, 85–8, 210
Pringle, Sir John, M.D., P.R.S., 21, 22, 45, 57, 61 ; notice of him, 55, 56
Puankhequa, Chinese merchant, 266
Pulteney, Sir William, 264

Q

Queen Caroline, of Naples, 198, 201, 202, 204, 205
Queen Caroline (wife of George II), 272
Queen Charlotte studies botany, 83, 193, 196, 304

R

Radnor, Earl, 208
Raffles, Sir Thomas S., 273, 282, 296
Raleigh Travellers' Club, 152, 153
" Ralph Robinson," 100
Rask, Erasmus, 250
Rawdon, Lord, 145
Ray, John, naturalist, 117
Ray, Martha, 43
Redinger, Curious Animals, 182
Reeves, John, horticulturist, 282
Regent's Park, 283
Reichard, —, geographer, 278
Reikjavik, Iceland, 33, 245, 250, 252

Rennell, Major James, 151, 282, .287, 291
Rennie, John, engineer, 278, 311
Revesby, Lincs, 3, 4, 8, 12, 37, 69, 304-17, 323
Reynolds, John, artist, 16
Reynolds, Sir Joshua, 160
van Rheede, H., botanist, 117 n
Rio Janeiro, 17, 20, 21
Roberts, James, 16
Robertson, Dr. William, historian, 36
Robison, Dr. John, 162
Rodney, Lord, 122
Roentgen, 152
Romney Marsh, 104
Rosse, Lord, 287
de Rossel, Captain, 192
Rotterdam, Banks on visit there, 36
Rowlandson, T., 179
Roxburgh, Dr. William, botanist, xi, 114-16, 118
Roxburghshire, 55
Roy, General, 45, 79
Royal Academy, 178 n, 209
Royal Botanic Society, 283
Royal Geographical Society, 152, 294
Royal Horticultural Society, 260, 261, 317-20, 331
Royal Institution, 258-60, 300
Royal Society, 10, 12, 14, 15, 21, 39, 42, 48, 49, 50, 54, 55, 56, 57, 58-61, 72 et seq., 91, 97, 118, 120 125, 158, 170, 174, 178, 179, 191-2, 198, 207, 209, 211-12, 255-9, 281, 288, 292-3 301-2, 325, 326
Royal Society Club, 70, 90, 281
Rumford, Count, 210, 258, 259
Rumph, Georg E., botanist, 117
Russell, Alexander, botanist, 116
Russell, Claud, 116
Russell, John, R.A., 331
Russell, Dr. Patrick, botanist, 115, 116-20

S

St. Andrew's, Fifeshire, 55
St. Helena, 111, 112, 113, 129, 137, 159

St. John's, Newfoundland, 11
St. Paul's Cathedral, 35, 160; Churchyard, 85
St. Petersburg, 146
St. Vincent, W. Indies, 120, 121, 122, 137, 228, 229
Salisbury, R. A., horticulturist, 261
Salt, Henry, 283-6
Samulcotta, India, 114
Samwell, David, with Cook's last voyage, 53
Sandwich, John Montagu, fifth Earl, 9, 15, 22, 23, 25 n, 26 n, 41, 43 n, 49, 50, 215
Sandwich Islands, 144
Sandsinding, W. Africa, 152
Sardanapalus, quoted, 297
Sauer, Martin, 146
Savigny [? a cutler], 176
Savu, 17
Scarborough, 47
Scoffers and scoffing, 172 et seq.
Scoresby, Captain William, 282, 294
Scotland visited, 32, 34
Scott, George, draughtsman, 151, 152
Scott, Sir William, 160
Scurvy, Captain Cook and, 49
Seville, 106
Shaftesbury, Lord and Lady, 208
Shakespeare tavern, 119
Shannon River, 144
Sheep and wool, 102-8, 219-21, 308 et seq.
Sheerness, 25
Shepherd, Rev. Dr. R., 162
Sherborne, Lord, 105
Ships : Adventure, 27 n, 39, 40, 167 ; Amphion, 209 ; Assistance, 137, 140 ; Bounty, 40 n, 129, 130, 133-7 ; Bridgewater, 235 ; Cato, 235 ; Chatham, 144 ; Congo, 279 ; Crescent 151 ; Cumberland, 235, 236 ; Discovery, 49, 51, 53, 54, 143, 144, 146 ; Endeavour, 17, 20, 21, 28-9, 32, 65, 111 ; Géographe, 233 ; Guernsey, 35 ; Investigator, 230, 231, 233, 238 ; Lady Nelson, 232, 251 ; Margaret and Anne, 251, 252, 254 ; Niger, 10 ; Orion, 252,

254 ; *Otter*, 237 ; *Pandora*, 135 ; *Porpoise*, 226, 229, 233–5 ; *Prince of Wales*, 140, 141 ; *Princess Royal*, 140 ; *Providence*, 137 ; *Resolution*, 24–7, 39, 40, 44, 45, 49, 53, 54, 129, 146, 214 ; *Rising Sun*, 167 *n* ; *Sappho*, 251 ; *Sirius*, 216 ; *Supply*, 216 ; *Talbot*, 252, 253 ; *Union*, 103
Ships provided with gardens, 126 *et seq.*
Siam, 114 *n*, 115
Sibthorp, Humphrey, botanic professor at Oxford, 7, 8
Sibthorp, John, botanist, 7
Sierra Leone Company, 153, 155
Silbury Hill, Wilts, 13
Simpkin, of the *Crown and Sceptre*, 70
Sinclair, Sir John, 100, 105, 106, 107, 159, 308
Skalholt, Iceland, 33
Skard, Iceland, 33
Skelton Castle, Yorkshire, 48
Sloane, Sir Hans, his botanical collections, 13, 66, 98
Smeathman, Dr. Henry, 71
Smeaton, John, engineer, 160
Smith, Dr. Adam, 162
Smith, Christian, botanist, 279
Smith, Christopher, botanist, 138
Smith, James, of Norwich, 257
Smith, Sir James Edward, 212, 257, 299
Smith, John (of Kew), 145
Smith, J. T., 324
Smith, Dr. Robert, master of Trinity College, Cambridge, 8
Snart, —, 109
Soane, Sir John, 287
Society of Antiquaries, 158, 159, 315
Society of Arts, 10, 100 *n*, 120, 121, 122, 124, 125, 158
Society Islands, 128
Soho Square, 4, 38, 62, 64, 71, 90, 187, 192, 242, 257, 259, 287, 288, 291, 292, 316, 319, 324
Solander, Daniel C., 16, 17, 20–2, 27–30, 32–4, 37, 39, 45, 56–7, 62, 64–6, 97, 118, 174, 176, 213, 290
Solomon, King, 157
Somerset, Duke of, 258

Souter, Captain, 167 *n*
South Africa, 97
Spalding, Lincs, 313
Spalding Gentlemen's Society, 4, 313
Spanish wool, 102 *et seq.*, 219, 220, 310
Spilsby, Lincs, 313
Spithead, 130, 237
Sporing, Henry, draughtsman, 16, 20
Spring Grove, Heston, 91, 145, 261, 293, 297, 300, 305, 316–321, 330
Staffa, 33, 92
Stanhope, Rt. Hon. Edward, 306
Stanhope, Lady Hester, 282
Stanhope, J. Banks, 306
Stanfield, Lincs, 3
Staunton, Sir George T., 268, 270, 271–3, 292
Steele, Joshua, 124, 125, 138
Stephensen, Magnus, 245, 247, 250
Stephensen, Olaf, 245, 248, 250
Stillingfleet, Benjamin, botanist, 13
Stixwould, Lincs, 3
Straungiörde, Iceland, 33
Strawberry cultivation, 261, 321 *n*
Streatham, Surrey, 41
Stuart, Andrew, 145
Stukeley, Dr. William, 3, 315
Sturt, Humphrey, of Crichel, 13
Suffolk, 310
Sumatra, 274, 282, 290, 291
Sunda Straits, 129
Surinam, 168
Sussex, Duke of, 287
Sutton, Captain, 103, 104
Suttor, George, 4, 229, 328
Sydney, N.S. Wales, 213, 217, 226
Sydney, Lord, 126, 215

T

Tahiti [=Otaheite]
Tangier, 216
Tankerville, Lady, 83 *n*
Tarrant Gunville, Dorset, 12
Tasmania, 253
Tattershall, Lincs, 3
Taunton, Somerset, 13
Telescope, Herschel's, 63, 64

Thames River, near Chelsea, 6
Thames and Severn Canal, 99
Thompson, Benjamin [=Count Rumford]
Thompson, R., 104
Thunberg, C. P., botanist, 65, 97
Tibet, 270, 298
Tierra del Fuego, 17, 19
Timbuctoo, 152, 284
Timor, 134, 233, 236, 241
Tobago, W. Indies, 120
Tobermory, 33
Tomkinson, Captain, 237
Torres Straits, 17, 137
Tothill Fields prison, 253
Totnes, Devon, 3
Townley, Charles, 38
Trampe, Count, 252
Tranquebar, 45, 115, 116
Transit of Venus, 1769, 14, 17, 20
Tripoli, N. Africa, 147, 149
von Troil, Uno, 32, 33, 68 ; his *Resa till Island*, 34
Truro, 177
Tuckey, Lieutenant J. K., 278, 302
Tupia, Otaheitan native, 20
Turner, Dawson, ix
Turtle story, 70
Tyrrell, —, 45
Tyrwhitt, Sir Thomas, 265
Tyson, Rev. Michael, 43, 57
Tyssen, —, 306

U

Ulloa, naturalist, 195, 196
Upsala, Sweden, 32, 34, 97 *n*, 155

V

Valentia, Lord (=Mountnorris), 283, 284, 286
Vancouver, George, 40, 143–5
Van Diemen's Land (=Tasmania) 230, 236, 240, 242
Vauxhall songs, 70
Vidoe, Iceland, 250
Vine culture in Australia, 229
Visiting tickets, collection of, 198, 323
Vizagapatam, 116

W

Walcot, packet agent, 103
Walden, —, 26
Wales, tour in, 37
Wales, William, astronomer with Cook's second expedition, 27, 40
Wallen, Matthew, horticulturist, 122, 123
Wallich, Nathaniel, botanist, 274, 282
Walpole, Horace, 173, 174, 175
Walter, John, founder of *The Times*, 166
Wardle, Colonel, 107
Warren, C., 77
Waterton, Charles, naturalist, 301, 302
Watson, Richard, Bishop of Llandaff, 145
Watson, William, M.D., 10, 63
Watt, James, engineer, 191, 278
Wedgwood, John, 260, 261
Weld, C. R., quoted, 15, 56, 73, 78, 79, 255, 260
Wells, Somerset, 13
West Indies, 97, 126, 129, 137, 145 ; botanic gardens there, 120 *et seq.*, 138, 159
West, James, P.R.S., 10
Westall, William, artist, 230, 235
Weymouth, 108
Wharton, Admiral Sir W. J. L., 31
Whewell, Dr. William, F.R.S., quoted, 59
Whitby, 54
White, Rev. Gilbert, 16
Wilberforce, William, 153, 154
Wiles, James, botanist, 138
Willis, Browne, antiquary, 3, 4
Willughby, F., naturalist, 117
Wilna, 171
Wilson, H. H., 287
Wiltshire, 105
Wimbledon, Surrey, 277, 319
Winchilsea, Earl of, 259
von Windischgratz, Count Joseph Nicolas, 83, 85
Windham, Rt. Hon. William, 160, 212
Windsor, 64, 67, 91, 109, 272, 311
Winterton, Lincs, 313
Witham River, Lincs, 312
Withering, William, botanist, 225

Wollaston, Dr. William H., 78, 301
Wolcot, John, M.D. [Peter Pindar], 177, 178
Wood, Sir Mark, 264
Woodford, E. A., 212
Wool, 99, 102 *et seq.*, 219–21 309 *et seq.*, 320
Woolwich, 35, 74, 292
Worsley, Sir R., 38
Wright, Dr. William, 247

Y

Yakutsk, 146
Yarmouth, 247

Yonge, Sir George, 113, 121, 136, 215
York, 46, 314
Yorke, Sir Charles, 238, 285
Young, Arthur, 100, 159, 308–10, 316
Young, Dr. George, horticulturist, 120, 121, 122

Z

Zaire River [=Congo], 279
Zimmermann, Henry, 53
Zoffany, J., 24, 25

Printed in the United States
By Bookmasters